Bernard Schmidt developed the Schmidt telescope based on the innovative idea of using a spherical mirror instead of a parabolic one. The Schmidt system represented a decisive advance in the development of the technology of astronomical telescopes. *Astrophotography with the Schmidt Telescope* reviews the history and achievements of this type of telescope, which has become so important. It includes reproductions of 43 photographic plates produced by large Schmidt telescopes, accompanied by explanatory text. The book will be of equal value to professional and amateur astronomers.

ASTROPHOTOGRAPHY WITH THE SCHMIDT TELESCOPE

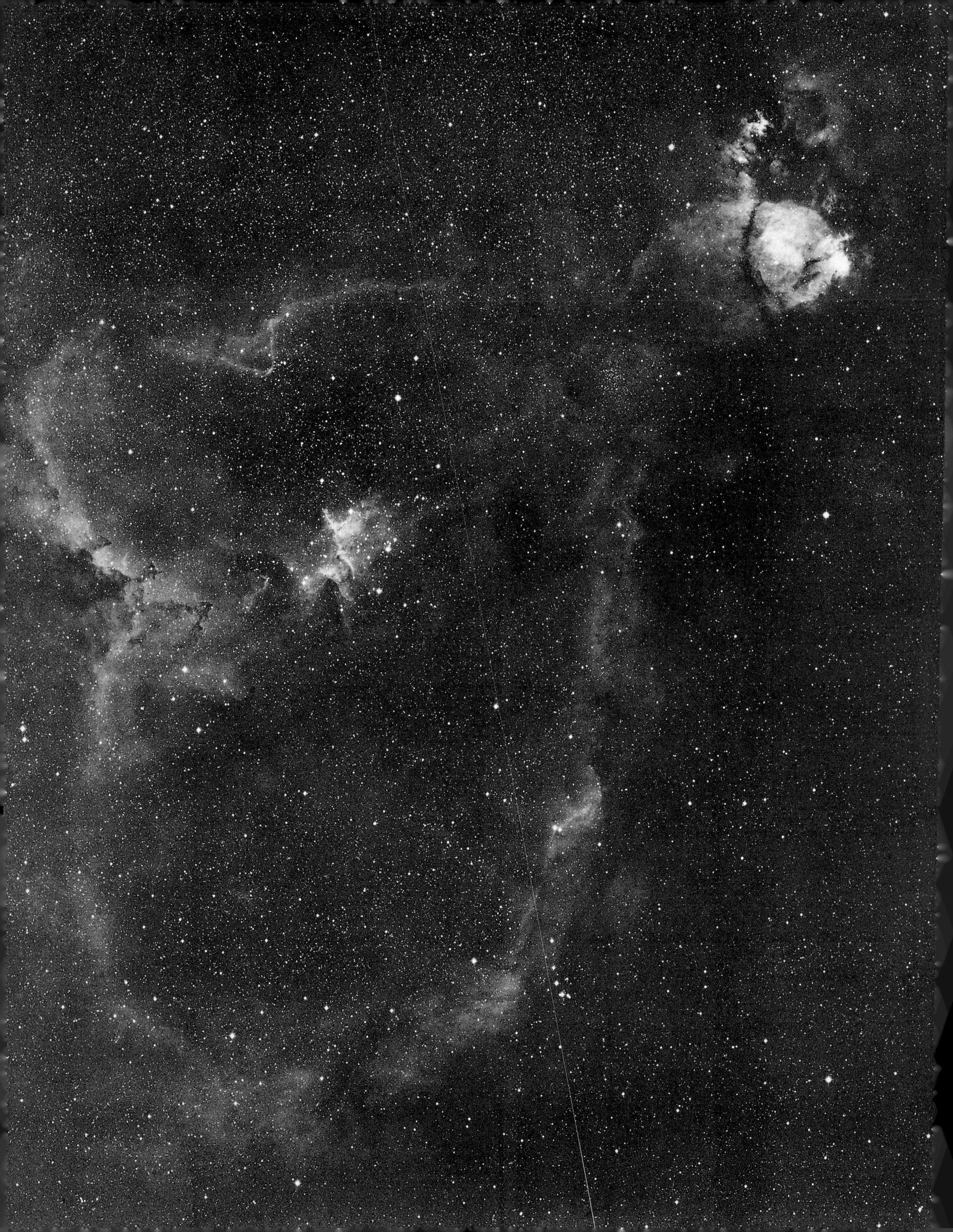

ASTROPHOTOGRAPHY WITH THE SCHMIDT TELESCOPE

S. Marx
Karl Schwarzschild Observatory, Tautenburg

W. Pfau
University Observatory of Friedrich-Schiller
University of Jena

Translated by P. Lamble

CAMBRIDGE
UNIVERSITY PRESS

Published by the Press Syndicate of the University of Cambridge
The Pitt Building, Trumpington Street, Cambridge CB2 1RP
40 West 20th Street, New York, NY 10011-4211, USA
10 Stamford Road, Oakleigh, Victoria 3166, Australia

Originally published in German by Urania-Verlag as
Himmelsfotografie mit Schmidt-Telescopen © Urania-Verlag,
Leipzig-Jena-Berlin 1990

First published in English by Cambridge University Press, 1992
as *Astrophotography with the Schmidt Telescope*

Printed in Great Britain at the University Press, Cambridge

A catalogue record of this book is available from The British Library

Library of Congress cataloguing in publication data available

ISBN 0 521 39549 6 hardback

Frontispiece:
The galactic emission nebula IC 1848 in Cassiopeia.
During the exposure, an artificial satellite crossed the
field of view. (Karl Schwarzschild Observatory,
Tautenburg)

WV

Contents

Preface

Photographs taken by E. E. Barnard and M. Wolf before the turn of the century played an important part in the development of people's conception of the appearance of the sky beyond the range of the naked eye. Neither of them used telescopes to take their pictures, but instead employed small portrait lenses of the type used in professional photography. Correspondingly, the pictures are notable for the large areas of the celestial sphere that they cover. The Schmidt Telescope, built by Bernhard Schmidt at Hamburg Observatory in 1930, emulated this early type of astrophotography to a certain extent.

Since – for the time being at least – a climax has been reached in the astronomical application of this optical system with the employment of the largest Schmidt telescopes, it is worth casting a look back at the history and achievements of this type of telescope, which has become so important.

This book includes reproductions of 43 photographic plates produced by large Schmidt telescopes, mainly at the Karl Schwarzschild Observatory in Tautenburg. This book stands out from earlier works of this kind because of the way in which the complete information content of the original plates has been retained in the process of photographic reproduction. New darkroom techniques played their part in this process. W. Högner, chief photographer at the Karl Schwarzschild Observatory has been able to develop these after many years of experimentation.

The positions of the objects depicted on the sky (or more precisely their coordinates in right ascension) were used to decide the order in which they should appear in the book. The coordinates are based on Epoch 2000.0.

In order to give the pictures some sort of framework, and in an attempt to underline the scientific significance of the objects, each picture is accompanied by an explanatory text. Other chapters of this book deal with the role played by astronomical research among the natural sciences; with technical details of the hardware and its history; and with the life of Bernhard Schmidt himself.

Our aim was to produce this book in a generally understandable form, so that each group of readers can gain some benefit from it – whether from the factual information presented or by simply taking pleasure in the illustrations of celestial objects. This book was never intended to be a textbook so it was not, therefore – especially in the illustrated section – arranged to explain specific points systematically. Readers whose knowledge of astronomy is not very extensive will be able to make good use of the Glossary. The units of measurement employed are those that are normally used in astronomy. Again, the necessary explanations can be found in the Glossary.

We would like to acknowledge our thanks to the many unnamed workers at the Karl-Schwarzschild Observatory who, in the course of their observational activities took the photographs from which we chose the selection presented in this book. We would especially like to thank W. Högner at Tautenburg, without whose cooperation and excellent work this book would not have been possible. We would also like to thank Dr D. F. Malin of the Anglo-Australian Observatory and Dr R. M. West of the European Southern Observatory (ESO), who made available to us pictures taken using the large Schmidt telescopes at those institutions. We owe our thanks too to Prof. Dr K.-H. Schmidt, Babelsberg Observatory and Dr W. Götz, Sonneberg Observatory who with their expert advice made important contributions towards the completion of our manuscript. Last, but not least, we would especially like to thank the publishers for their understanding and patience.

Siegfried Marx and Werner Pfau
Tautenburg and Jena

1 Astronomy as a natural science

Even in prehistoric times the starry sky was an object of observation and reflection. Rock carvings and the alignments of megaliths that have astronomical orientations may be regarded as the first records and show how great a role astronomy played in the spiritual life of the earliest cultures. Given the high esteem in which knowledge was held, and the level of knowledge that was reached by the Babylonian and Greek cultures – especially concerning the motion of the Sun, the Moon and the planets – one can justifiably describe knowledge related to the stars as the oldest branch of natural science.

Astronomy is a fundamental part of our natural sciences today. Through close interaction with other disciplines, above all with mathematics and physics, it has remained an expanding science, which on the one hand triggers off advancements in knowledge, and on the other profits from them. Results from almost all branches of astronomy provide possibilities for testing various aspects of physics but at the same time also raise many challenges for physics. Here one might consider the structure of the universe as a whole, which is a testing ground for various gravitational theories; or the plasma processes in the outer layers of individual stars where magnetohydrodynamic theories are subject to verification. The great majority of astronomical objects are characterized by large masses and enormous energies. These generally cause matter to exist under extreme conditions of temperature and density, which, even today, still lie well beyond what can be realized in a physicist's laboratory. Study of astronomical objects therefore enables us to examine physical theories of the nature of the universe that range far beyond the conditions that we can experience here on Earth. On the other hand, astronomy is inconceivable without the application of the framework of knowledge provided by physics – at least since the time of Kepler, who was the first to raise questions about the physical causes of the motion of the planets. Astronomical research is 'alive' and can only develop with the application of physical laws, theories and data. The same holds for the mathematical procedures that are indispensable for the solution of theoretical and other problems.

A general aim of natural scientific research is to recognize and understand in a comprehensive way our material world with its phenomena, inner relationships and obedience to the laws of nature, and its evolutionary processes. These efforts come, on the one hand under the aspect of the practical and useful application of science for society, but also have a considerable spiritual component, and thus enable us to form a factually based conception of the universe. In this respect, astronomy has always played a decisive role. Newton derived his law of gravity by linking the law of falling bodies, established on Earth by Galileo, with the laws of planetary motion discovered by Kepler. For him, the epistemological basis for this was the uniformity of physics on Earth and in the universe as a whole.

With the consolidation of knowledge that was triggered off by increased practical applications and the formulation of specific methods of gaining knowledge, the entire system of science fanned out at the start of the modern age into relatively independent disciplines each with more or less clearly defined research topics. This process, like general social development, evolved so rapidly at times that smaller, individual subject-areas developed and became independent. One could take as an example of the new physics the split of quantum physics away from classical physics that took place just after the turn of the century.

The various forms of matter that we encounter away from the Earth at great distances in space and time are the subjects of our astronomical research. The evolution of individual objects and their motions with time therefore represent a central question in modern astronomy. Scientists are concerned with understanding the observed phenomena as the result of evolutionary processes and with tracing them back to their initial phases. Both the size of the object and the predominating concept of evolution are of permanent relevance to the world-picture provided by the results of astronomical research.

Astronomy left its early scientific stage with Johannes Kepler, who described the motions of the planets as being in accordance with specific laws and introduced the concept of a force as the cause of the observed motion. The title of Kepler's work, published in 1609, *Astronomia nova, αιτιολογητος seu physica coelestis* ['*New Astronomy*, causally explained or *Celestial Physics*'], reveals the new approach (Fig. 1.1). When analysing the positions of the planet Mars that had been measured by Tycho Brahe,

Fig. 1.1. Title page of the first edition of Kepler's Astronomia Nova, *published by E. Vögelin in Heidelberg.*

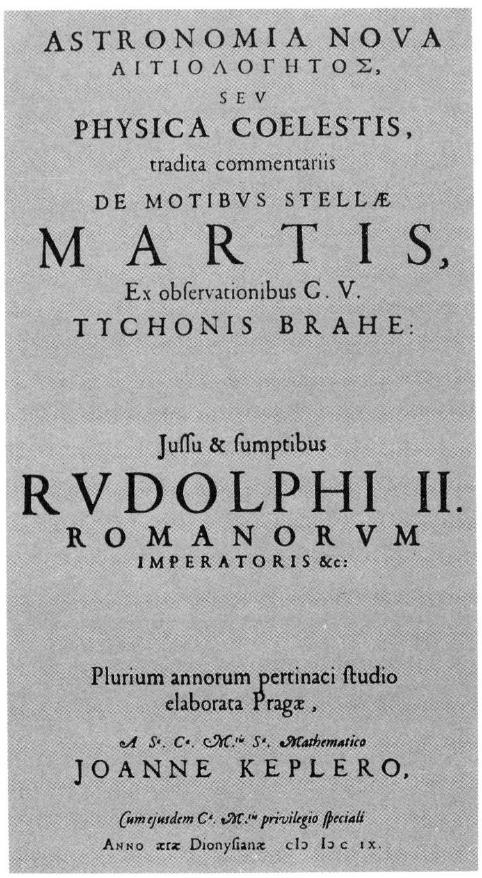

ASTRONOMIA NOVA
ΑΙΤΙΟΛΟΓΗΤΟΣ,
SEU
PHYSICA COELESTIS,
tradita commentariis
DE MOTIBVS STELLÆ
MARTIS,
Ex obſervationibus G. V.
TYCHONIS BRAHE:

Juſſu & ſumptibus
RVDOLPHI II.
ROMANORVM
IMPERATORIS &c:

Plurium annorum pertinaci ſtudio
elaborata Pragæ,

A S. C. M.tis S. Mathematico
JOANNE KEPLERO,

Cum ejusdem C. M.tis privilegio ſpeciali
ANNO æræ Dionyſianæ clɔ Iɔ c IX.

Kepler freed himself from dogmatic prejudices and restrictive theories but, at the same time, he accepted the criterion that a theory should agree with the observational data, within the limits set by the accuracy of the measurements, which had to be carefully assessed. Until then the collection of facts and the purely descriptive representation of observations had been the primary concern. Even Copernicus with his new heliocentric view, had interpreted the observed planetary positions on the basis of Aristotle's 'natural' forms of motion – circular movements and uniform velocity. Nevertheless, the fact that his work dethroned the Earth as the centre of the planetary system was of the greatest ideological and philosophical significance, and effected a 'Copernican Revolution' in scientific thought. The link led from Copernicus via Kepler and Galileo directly to Isaac Newton (Fig. 1.2). As mentioned above, he was successful in deriving his law of universal gravitation from motions observed in the planetary system, especially from those of the Moon. The law of gravity has become a fundamental law of physics and of all natural science. With it as a basis, and with constant re-examination of, on the one hand, the changes in the positions of the planets, which were continually being measured more accurately, and, on the other, of mathematical methods, celestial mechanics achieved its first full flowering. This high point was marked by the five-volume *Méchanique Céleste* of Pierre Simon Laplace, published between 1799 and 1825.

Celestial mechanics was instrumental in promoting mechanics as a branch of physics. With the almost simultaneous discovery, by three independent observers, of the first stellar parallax (in 1838), the periodic motion of 'fixed' stars that had been postulated by Copernicus as a necessary consequence of his ideas, reflecting the orbital motion of the Earth around the Sun, was at last confirmed. A few years later, from perturbations in the orbit of the planet Uranus and with complete confidence in the validity of the law of gravitation and in the methods of celestial mechanics, people were able, on purely theoretical grounds, to infer the existence of another planet. This planet, Neptune, was discovered in 1846, very close to the calculated position. Alongside such triumphs of positional astronomy and celestial mechanics, other fields of astronomical research remained in the shade. It was therefore not surprising that F. W. Bessel, leading representative of this successful field, saw the actual task of astronomy as being '. . . laying down rules, by which the motions of the heavenly bodies, as they appear to us on Earth, could be determined'. Everything else, i.e. what we now classify as astrophysics or planetology, was '. . . not unworthy of attention, but does not affect the true interest of astronomy'.

Fig. 1.2. Isaac Newton (1643–1727).

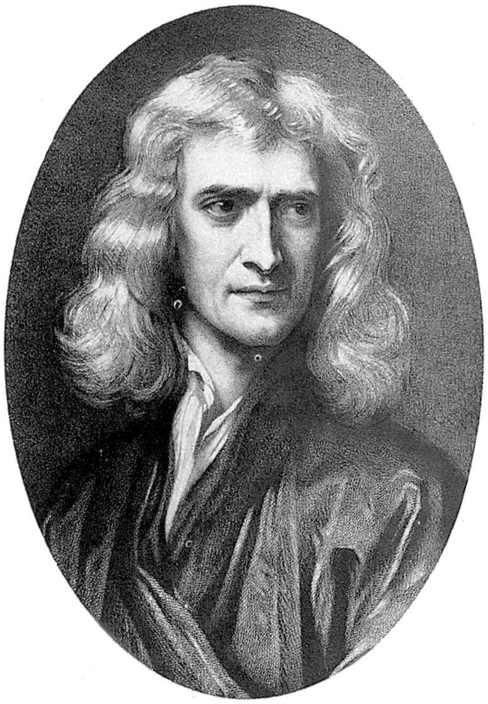

Astrophysics – one branch of astronomy

The incorporation of new physical procedures and ways of thought that go beyond those in mechanics, that has increasingly taken place since the middle of last century, led to the development of astrophysics as a branch of astronomy. Because of its close interaction with physics, it developed very successfully, and quickly became the leading field in astronomical research.

One of the roots of astrophysics lies in the discussion that centres around the structure of the universe. After the shock of the action of the Papal Inquisition against Giordano Bruno, who was burnt at the stake in 1600, in part because he had taught that the universe was infinite and that an infinite number of suns existed, and also against Galileo as the advocate of Copernican teachings in the trial in 1633, discussion on this point remained silent for over a hundred years. It was then revived, quite suddenly, with the works of Immanuel Kant (Fig. 1.3). Stimulated by the strongly theologically inspired ideas expressed by Thomas Wright, in his book *Allgemeine Naturgeschichte und Theorie des Himmels* ['*General Natural History and Theory of the Heavens*'] (1755), Kant explained our Galaxy as having been created mechanically 'according to Newtonian principles'. His conclusions concerning the extent of the system, its flattening by overall rotation and the nature of that rotation seem to be very modern. He also considered the Galaxy to be a member of a hierarchy, lying between the much smaller Solar System on the one hand, and a grouping of numerous related galaxies on the other. This theoretical idea, deduced from the law of gravity, was then pursued by William Herschel in his attempt to 'gauge' the extent of the Galaxy by counting stars visible through his telescope. Herschel also saw the small nebulous spots as being other galaxies similar to our own – at least, that is, until other observations made him uncertain. He listed them incidentally, according to their appearance, in an evolutionary sequence. The large telescopes of our century then produced results that were to follow on from Kant's and Herschel's starting points in such a splendid way. By now they have extended observable space so far that radiation from the most distant sources has taken some 15 thousand million years to reach us.

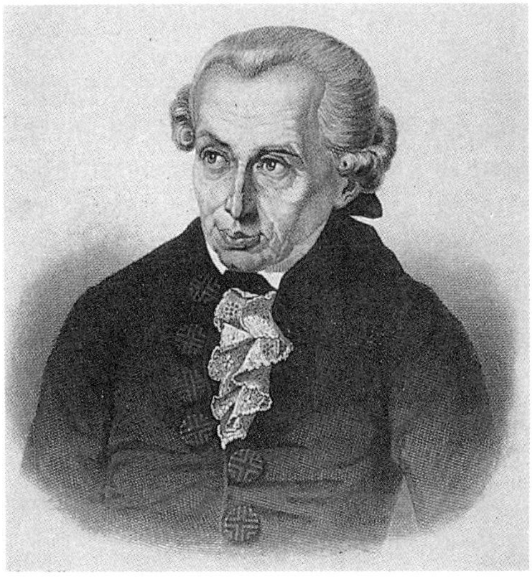

Fig. 1.3. Immanuel Kant (1724–1804).

This early, close connection between astronomy and physics is more obvious in the field of spectroscopic solar and stellar research. By examining glowing gases G. Kirchhoff and R. Bunsen had established that, according to the experimental arrangement, the gases either emitted or absorbed radiation in sharp spectral lines, and that the appearance of these spectra is characteristic of the individual chemical elements. Kirchhoff then described his experiments in his now-famous work on the Sun: *Untersuchungen über das Sonnenspektrum und die Spektren der chemische Elemente* [*Investigations of the Solar Spectrum and the Spectra of the Chemical Elements*] (1861). Assuming that the physics at work in the laboratory and the Sun were one and the same, he concluded that glowing gaseous matter was the source of the Sun's radiation and that a series of familiar chemical elements occur in these hot layers. A method of research had been born which, from the detailed interpretation of the spectra of celestial objects, allows conclusions to be drawn about their physical state, chemical composition, motions, magnetic fields and other characteristics. Together with the comprehensive application of the laws of physics, spectroscopy has come to be the basis of modern astrophysics.

The fact that the universe is subject to evolution, in other words that it has a history – which is very important for our conception of the universe – can be regarded as a general result of astronomical research in the twentieth century. This evolution can be seen in individual stars, as well as in galaxies and in the universe as a whole. The life of a star is now accurately known from its beginnings through to the various later stages. There are also well-established ideas on the formation of galaxies and their evolution through both active and quiescent phases.

At the beginning of the 1920s,

various authors independently deduced from Einstein's field equations that the whole universe might be subject to evolution with time. This was confirmed when proof of the overall expansion of space was obtained – by Edwin Hubble in particular. At present, questions about the initial phases of the universe are highly topical and are discussed in terms of highly complex quantum physics and the physics of fundamental particles. Fundamental developments are taking place in astronomy over the entire range of investigation, from late stages of stellar evolution to cosmology.

The fundamental means by which an astronomer gains access to the objects of his research is electromagnetic radiation. The history of astronomical instrumentation has therefore been characterized by the attempt to use the quanta of radiation that are received more and more effectively as sources of information. The various measuring techniques are aimed at analysing the spatial, temporal and spectral distribution of the photons emitted by celestial objects. To these we may add the determination of the polarization and the phase-relationships of the radiation. Ideally, the result is a complete picture of an object's position, extent and three-dimensional structure and the characteristics of the radiation received. Any factors may be variable over a broad range of different time-scales. Additional information of this type always proves to be significant in connection with the physical nature of the objects.

Progress through new observational techniques

The history of astronomy also offers many examples of the fact that the introduction of new working techniques leads to decisive advances in knowledge. The new problems thus raised often stimulated technical aspects of measurement and analysis. Examples can be found in the telescopes themselves, which, either through appropriate adaptation for specific purposes or through previously unattainable accuracy and measurements, produced new results. They could just as well be taken from the field of auxiliary instruments used on telescopes, such as photometers, spectrographs, and the radiation-detectors they employ. Here we need only refer to photographic methods. Since the introduction of these methods at the end of the last century, it has been possible to collect a truly indescribable amount of material, thereby opening an effective way to investigate space outside the Solar System (Fig. 1.4). Only brief mention need be made of the fact that the development of new opportunities for carrying out measurements – both from the ground and from satellites – have triggered off an explosion of knowledge in spectral regions outside the optical range. Continual innovations in methods of processing data have proved to be particularly significant. Such data are today obtained with electronic detectors mounted directly on the telescopes, or arise as the results of analysing photographic plates. We include here the mathematical and computational methods involved in modern data- and image-processing, which enable the all-important signals to be differentiated from the many natural and instrumental forms of degradation, thus ensuring that the information content is available for use.

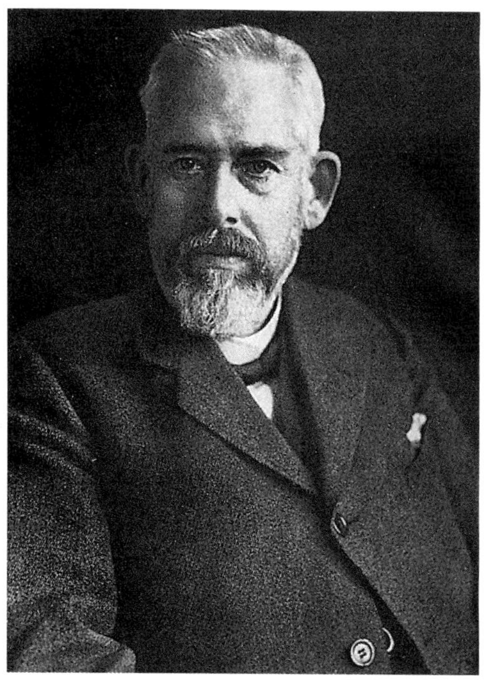

Fig. 1.4. Max Wolf (1863–1932) became a pioneer in astrophotography at Heidelberg and remains one of the most successful astronomers of his time.

The Schmidt telescope is also a typical example of how new technical developments can trigger off scientific progress. The Schmidt telescope combines large aperture, favourable ratio of aperture to focal length, and extraordinarily wide field of view with a highly efficient photographic process. Thus a vast amount of information can be obtained in each photograph, and can be stored for later analysis. Schmidt exposures are therefore especially suitable for the examination of extended objects and for statistical programmes that require large samples of stars and extragalactic systems.

In astronomy, observation provides both the fundamental material on which scientific hypotheses and theories are based, and the criteria by which their validity is judged. In this respect, it differs from other natural sciences, such as physics, which are largely based on experiments that directly affect the objects being investigated. The modern astronomer's observational material consists exclusively of data gained objectively with the help of sophisticated instrumentation. The data should be processed in stages so that the specific information required by any particular investigation should emerge as clearly as possible. Together with other results, the data may then enable us to deduce more general correlations. Such concepts have to be verified, by testing their agreement with the existing system of knowledge, by new measurements, or by some other suitable means. Determining suitable controls for the hypothesis that is to be tested, choosing an appropriate process for testing it, and specifying the data to be obtained – sometimes by employing new methods as well – are all important parts of scientific work. This is how astronomers experiment with the wealth of individual data that is available to them. They compare their ideas with observational findings and alter, as far as possible, the initial parameters until an adequate agreement between model and reality is reached, or until the starting hypothesis is proved to be untenable and has to be fundamentally altered.

Every possible form of physical condition and evolutionary state that can be found in the universe is represented among the immense number and diversity of objects in the sky. The various phenomena that the experimental – in the true sense of the word – scientists try to produce by directly acting on the subjects of their investigations, are, so to speak, laid out simultaneously alongside one another on the celestial sphere in front of astronomers' eyes. Naturally, for purely practical reasons, not all the characteristic phenomena are initially as accessible as one might wish. But it is just this that, in astronomy just as in any other experimental discipline, provides the challenge for further development of instrumentation and methods. This is what leads to scientific progress. The brilliant successes of astronomy in the twentieth century are indeed based on the wider range of optical telescopes that have been used to study faint objects and, in particular, on the enormous expansion of the portion of the whole electromagnetic spectrum that can be covered by our instrumentation. The latter has resulted from the rapid physical and technical developments that have taken place since the beginning of the 1950s. As a result, completely new astronomical objects and classes of objects have become accessible. Apparently well-known phenomena have revealed unexpected facets and have provided unexpected insights. Previously unsuspected conditions and processes have also been detected and require interpretation. Our view of the structure and evolution of the universe has, in recent years, been drastically expanded. For the time being, no end can be seen to this process.

2 The construction of astronomical telescopes

The development of astronomical telescopes from their beginnings up to their current state is marked by many different stages and names, and also involves a lot of detail. This development had many ramifications and led from the first telescopes, which were used purely for simple observation and exploration of the heavens, to the current large instruments employed in making measurements that form the whole basis of modern astronomy. One of the most recent of those ramifications was the Schmidt telescope. The development was supported by the general progress in technology, by craftsmanship, by specific requirements and by brilliant ideas.

The telescope in the form of the refractor was most probably invented in 1608 by Hans Lippershey, a spectacle-maker in Middelburg in Holland. Situated near the mouth of the river Schelde, the town was a flourishing centre for trade and crafts and, amongst other things, glass was produced there. It is remarkable how quickly the provincial administration and the Estates General in The Hague responded to Lippershey's application for a patent on his invention, and how quickly the telescope then spread throughout Europe. A report on Lippershey's application is dated 2 October 1608 and, within the following two months, delegates from the provinces had examined the instrument, called for a modification for more comfortable viewing with both eyes, and ordered more copies of this obviously satisfactory variant. The patent, however, was not granted, the comment being that 'the art of looking into the distance' had in the meantime become known in other towns. This may well have been true because, within the next few months, numerous reports were made about telescopes that had made their way into the hands of high-ranking persons in the form of presents. In the spring of 1609 telescopes are reported to have been for sale in two shops in Paris.

As a result of this very rapid spread, it is very probable that Galileo Galilei in Florence also heard about the invention and that this was enough to prompt him to make some experiments of his own. The first instrument that he produced had an objective with a diameter of 42 mm, a focal length of 2.4 m, and a magnification of nine times. A practical person for his time, Galileo showed the telescope to the Doge of Venice and referred to the military advantage that the republic would gain from being able to recognize and estimate enemy numbers at sea or on land at a very early stage. This was on 21 August 1609. The dimensions of several of the telescopes produced by Galileo have been recorded. They do not differ very much from those mentioned above but eventually a magnification of 30 times was achieved.

Galileo owes his prominent role in the history of the telescope to the fact that he was the first to use the new device extensively for astronomical observation, and because he radically extended the limits of the observable universe and opened up completely new fields with his discoveries about the planets and their satellites, and also about the stars.

Galileo's chosen combination of a convex objective lens and a concave eyepiece lens has found no lasting use in astronomy because of its small field of vision and unsuitability for measurement purposes. The telescope used in astronomy owes far more to the work of Kepler. Although we may mainly associate Kepler's name with the laws of planetary motion, at least a quarter of his writings are dedicated to optics. These include the work *Dioptrice* (Augsburg, 1611), which in a clearly organized manner describes the optics of refraction, the human eye, the effects of a simple lens and use of different combinations of lenses. It also contains rules for making the corresponding measurements and calculations. Among the lens combinations, first place is given to the description of the telescope – now named after Kepler – that is constructed from biconvex lenses. The theory of Galileo's telescope follows later. It was Galileo and his telescope that, in August 1610, had inspired Kepler to become involved with the theoretical basis of optics. This led to his book *Dioptrice* and ultimately to the existence of the whole scientific field of geometrical optics. The first practical implementation of a Keplerian telescope can be traced to Christoph Scheiner, who, with such an instrument, made his long series of observations of sunspots.

The quality of optical lenses in Galileo's day – even though he took his glass from the exceptional, Venetian, glass workshops – was greatly limited by the impurity of the material. For nearly the next 200 years, all of his successors had this same problem to contend with, until the introduction of new optical construction principles and of a new glass technology. At first one could only reduce the optical aberrations by stopping down the lenses to smaller apertures, or by using long focal lengths, which meant the construction of very long telescopes. A crucial aberration in a simple lens is chromatic aberration. In order to make its influence tolerable, the focal lengths of such lenses have to be much larger than their diameters and entail extremely long telescopes with all the associated problems of mounting. Examples of such telescopes include the one belonging to Johannes Hevel ['Hevelius'] at Danzig (Fig. 2.1), which he used for his observation of features on the surface of the Moon, with a focal length of 49 m; and the telescopes (up to 44 m long) used by Giovanni Cassini to discover the division in Saturn's rings since named after him.

The invention of the achromatic objective represented a great advance on such unwieldly telescopes. The discovery involved combining lenses made of materials with different refractive indices. Such lenses were first successfully produced by John Dollond in England in 1759. One typical-sized telescope designed by him had a 10-cm aperture and a focal length of only 1 m, yet nevertheless allowed a magnification of 150 times.

A particular milestone in the development of refractive optics and the construction of telescopes is associated with the name of Joseph Fraunhofer. By a fortunate combination of circumstances he was able to enter a mechanical and optical workshop in Munich, where he revolutionized both the construction of instruments, and the production of glass.

The hallmark of both the optical instruments that were produced in quantity in his workshops, and of the individual special orders was an outstanding quality. One reason for this was that Fraunhofer also introduced a method of precisely measuring the refractive index of various types of glass, and from this, to calculate the paths of rays through lenses by means of trigonometry (e.g. Fig. 2.2). It was therefore possible for the optical aberrations of even off-axis rays to be kept very low. Newly introduced test procedures during manufacture complemented the theoretical work. The telescopes produced by Fraunhofer's workshop were also outstanding for an advanced mounting that guaranteed the stability of the telescope whilst it was being rotated around the axis that pointed towards the North Celestial Pole. This design, known as the 'German mounting' is still in use today. Such an equatorial mounting, so convenient for track-

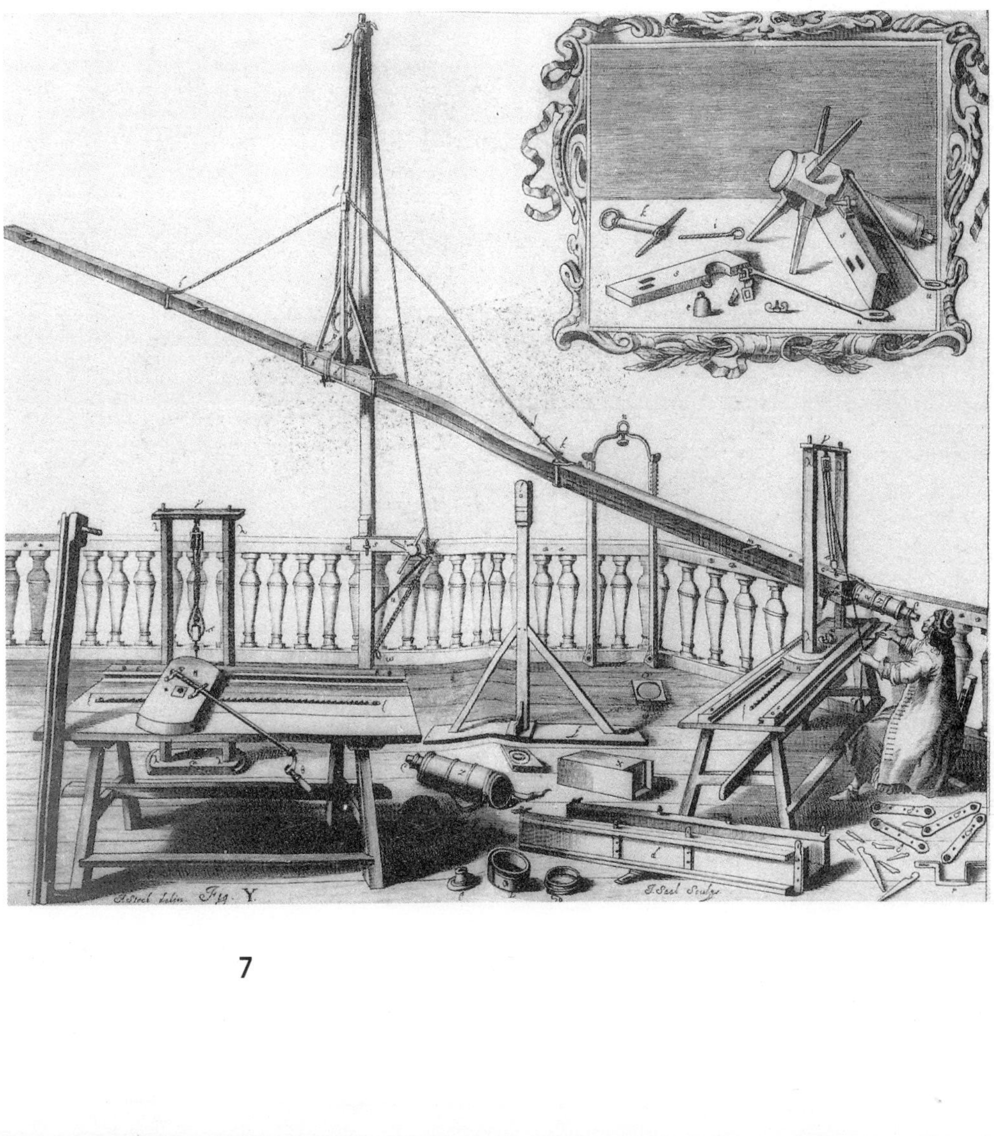

Fig. 2.1. Telescope of J. Hevelius (1611–1687) from his second main work Machina Coelestis.

ing astronomical objects during their daily motion across the sky, had been derived, before Fraunhofer's time, from the all-purpose instruments used for astronomical work, by tilting the azimuth axis at an appropriate angle to form a polar axis. As well as improving this type of mounting, Fraunhofer also introduced a clockwork drive to turn the telescope to compensate for the daily rotation of the sky.

The development of astronomical refractors reached a peak, but also came to an end, with large instruments of this type that appeared towards the end of the last century. One may mention here the Potsdam double refractor with its lenses of 80- and 50-cm diameter, corrected for photographic and visual observation respectively, or the largest refractor in the world, with an aperture of 1.02 m and a focal length of 19.4 m. The latter telescope was put into service in 1897 at the Yerkes Observatory of the University of Chicago.

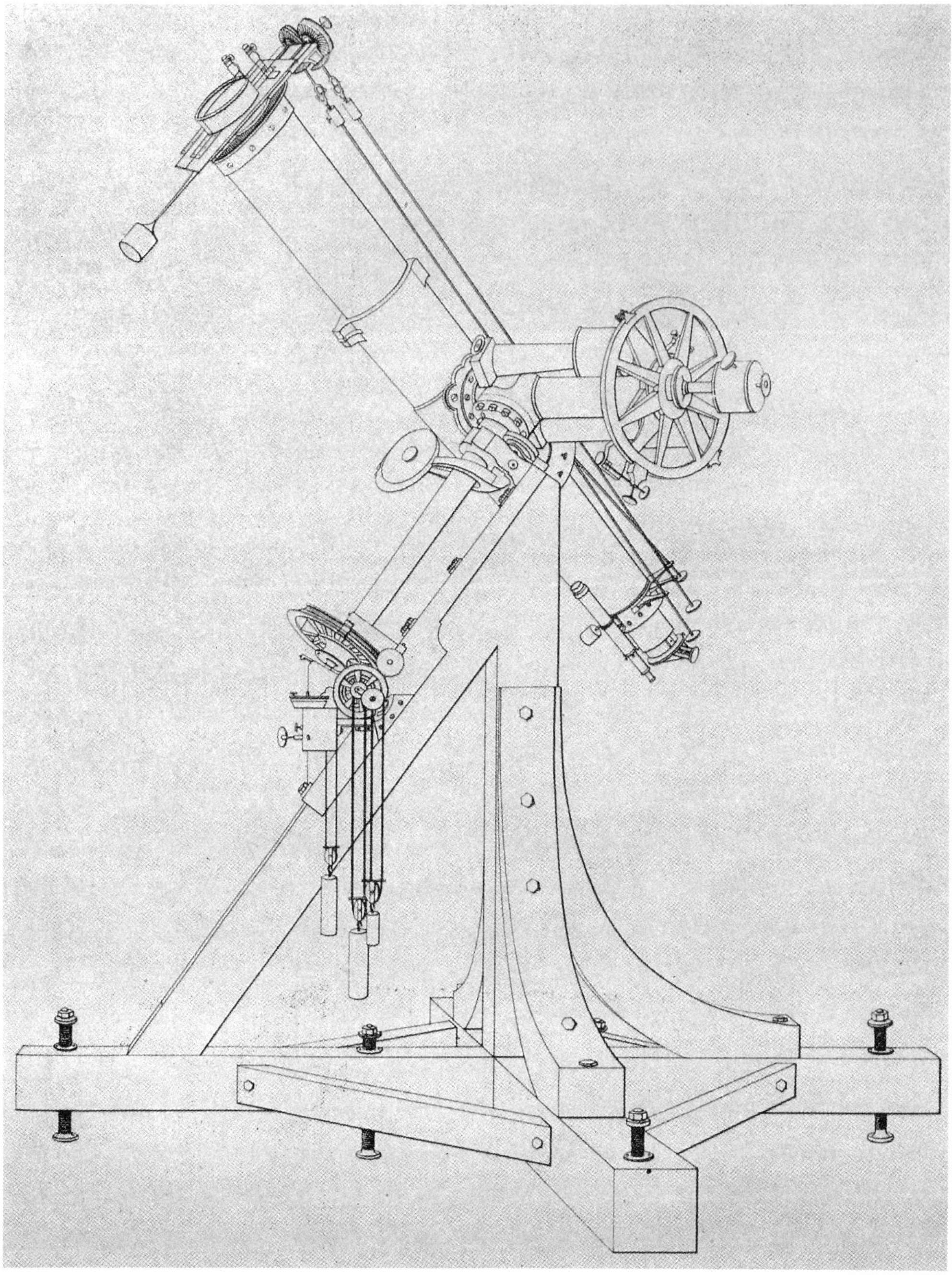

Fig. 2.2. The famous heliometer for the precise measurement of small angular distances, which F. W. Bessel commissioned J. v. Fraunhofer to make in 1824.

Historical reflecting telescopes

The development of astronomical reflecting telescopes has run in parallel to the progress made in refracting optics – sometimes being of secondary importance, then again taking the lead – until at the turn of the century it became preeminent. The introduction of reflecting telescopes is intimately associated with the name of Newton. Even though Galileo, Niccolo Zucchi (around 1616) and James Gregory (around 1660) had already voiced the idea of using curved, reflecting surfaces for optical imaging, Newton was the first to achieve a practical implementation of a reflecting telescope. After he had succeeded in casting a polishable metal alloy, Newton presented an instrument of this type with a 3.4 cm aperture to the Royal Society in 1672. He believed that, with glass lenses, the image quality obtainable was unavoidably limited by the chromatic aberration that occurs and therefore advocated reflective optics for aberration-free imaging. Newton used a small plane mirror placed in the light-path, in order to divert the pencil of rays reflected from the primary to the side of the tube and towards the eyepiece. By using concave or convex secondary mirrors, one obtains, respectively, the types of telescope named after Gregory or Cassegrain. It was Gregory who had suggested, in particular, that the rays should be carried through a central perforation in the primary mirror, along the direction of the optical axis, thus providing much more convenient handling; the directions in which the image was viewed and in which the telescope was pointed being the same.

For a long time the main technical problem lay in the production of mirrors with good, and above all, with lasting, reflective properties. Until the middle of the last century, different bright metal alloys were used, which had to be outstanding in their special surface characteristics. For example, a mixture of copper and tin, in a ratio by weight of approximately 2 : 1 was suitable.

Famous large telescopes from the era of metal mirrors were those built by William Herschel and later by Lord Rosse. Herschel, then living in Bath in the south of England, acquired the ability to produce telescope mirrors and the necessary tubes and mountings, in order to further his strong interest in astronomical observation. The exceptional optical quality of the instrument with which he discovered the planet Uranus in 1781, was confirmed by the then Astronomer Royal, Neville Maskelyne. By direct comparison Maskelyne found that Herschel's instrument was far superior to all those at the Royal Observatory at Greenwich. The annual allowance awarded by King George II, in recognition of the discovery of the planet, put Herschel in the position of being able to devote his life completely to astronomical work and, in particular, to the improvement of telescopes. To him, the latter meant, above all, the completion of larger instruments, and in 1783 he completed one with a 45-cm aperture and 6-m focal length. With this he was able to produce numerous astronomical results of lasting significance, because of the instrument's light-grasp. It clearly demonstrated to him, however, the great gain that could be achieved with a large aperture, so much so that he subsequently began construction of an instrument several times larger. Supported by further financial aid from the Crown, he started building the famous 40-foot telescope at Slough (Fig. 2.3). A tube more than 12 m long, carrying a metal mirror with a diameter of 1.20 m, was supported and positioned by a masive wooden construction, the erection of which at times required as many as 40 workers. The first test took place in 1787, but the mirror had insufficient mechanical stiffness against bending, and only with the third mirror casting was he able to achieve satisfactory images of the stars. An immediate success was achieved, which appeared to justify all the work involved. In the first month after operation finally began in 1789 Herschel discovered Enceladus and Mimas, the two inner satellites of Saturn. This, however, was to remain the only scientifically significant result that was to be achieved with the giant telescope. Its unwieldiness in operation, and unavoidable deformations of the mirror, caused by its mass of over 1000 kg or the results of thermal effects, could not be technically mastered at that time and hindered its practical use very considerably.

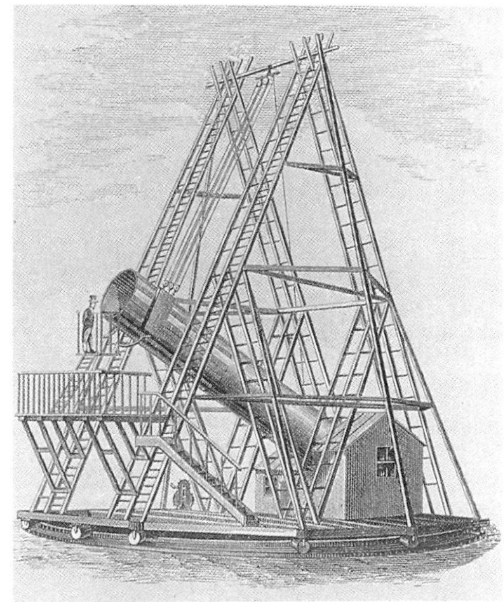

Fig. 2.3. The reflector with a focal length of 40 feet, completed in 1782 by William Herschel.

The optical arrangement of Herschel's large telescopes was such that observations were made at the direct focus of the primary mirror. The mirror was tilted slightly and the image was thus moved to the edge of the front aperture, so that it could be reached by the observer without causing a large amount of obstruction of the entering pencil of rays. With his smaller telescopes, Herschel followed Newton's type of construction.

In 1845, a telescope came into being that was to remain the largest until the opening of the 100-inch Hooker Telescope at Mount Wilson Observatory, i.e. for over 70 years. It was built by the third Earl of Rosse at his family residence, Birr Castle in central Ireland (Fig. 2.4). The dimensions justified the description 'The Leviathan of Parsonstown'. The total length amounted to 16.60 m (56 feet), the mirror diameter was 1.82 m (6 feet), and the mass of the mirror was 3.8 tonnes (this was the fifth casting despite many years of experiments to find the best metal alloy and most suitable polishing methods). The total weight of the tube was over ten tonnes. At that time such an instrument was only controllable by limiting its range of movement. The tube was supported on a specially constructed, lower swivel joint and was suspended at the front end by chains between two parallel, strong stone walls lying in the meridian with a height of approximately 15 m. This allowed movement between a northern zenith distance of 40° – i.e. the pole – and the southern horizon, with a lateral movement of 7.5° towards each side of the meridian so that objects under observation could be followed to a limited extent during their daily motion. This was later carried out by means of a clockwork device, which testifies to the ease with which the whole construction could be moved. The observer's access to the eyepiece was possible via one of several movable stages or ladders, according to actual zenith distance. The detailed drawings made by Lord Rosse, and later by his son, of galactic and extragalactic objects, and of the surfaces of planets, show the exceptional quality of the instrument. Lord Rosse was able, for example, to recognize, and illustrate for the first time, the spiral form of galaxies. Magnifications of up to 6000 times were possible, but atmospheric conditions limited the optimum power to 1300 times. The 'Leviathan' was also at the limit of then-current technology. According to a remark made by the builder himself, the metal mirror was not polished as often as necessary, simply because of the problem of dismantling the heavy mirror cell. The instrument was therefore often operated below its optimum efficiency. In order to solve the problem of light-loss with his metal mirrors, Herschel, for instance, always had several available and exchanged these – admittedly far more manoeuvrable – mirrors whenever necessary.

A decisive advance came in the middle of the last century, when it was learnt how to master the polishing of glass plates and how to deposit a thin layer of metallic silver onto the polished front surface, following the method developed by the chemist, Justus von Liebig (1856). A fresh layer of silver had a high reflectivity of approximately 95% in the visible spectral region but this fairly quickly falls to 70% and below as a result of tarnishing. For this reason, people subsequently switched to aluminium as a coating for mirrors. This is vaporized under a vacuum and, as a protection against mechanical damage, the deposit is covered with an additional thin, transparent layer of quartz.

Fig. 2.4. Lord Rosse's giant telescope.

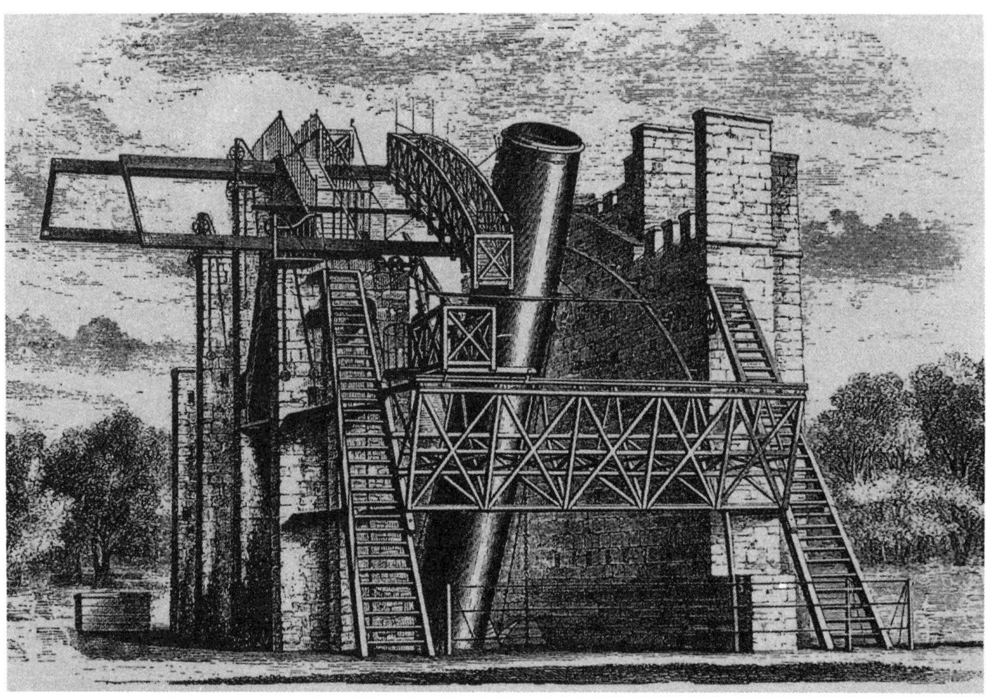

Newton had already made experiments with small glass mirrors. He had, however, tried to use the rear face of the glass as the reflecting surface and was halted by the bad quality of his glass in his day, because, with this arrangement, the light had to pass twice through the glass, giving a correspondingly low image quality. The first large glass mirror (with a diameter of 80 cm) coated with silver was made by the physicist and astronomer Léon Foucault (*circa* 1865).

Apart from Lord Rosse's giant telescope others dating from the second half of the last century should be mentioned because of their size and light-grasp. They all embodied new concepts, which were either the subject of further technical development or were reintroduced later. Foucault designed the 1.20-m reflector for the Paris Observatory in 1875. William Lassell worked with a telescope of the same dimensions on Malta in 1860 (Fig. 2.5). Both telescopes were mounted equatorially and were operated without a dome. The observer could reach the eyepiece

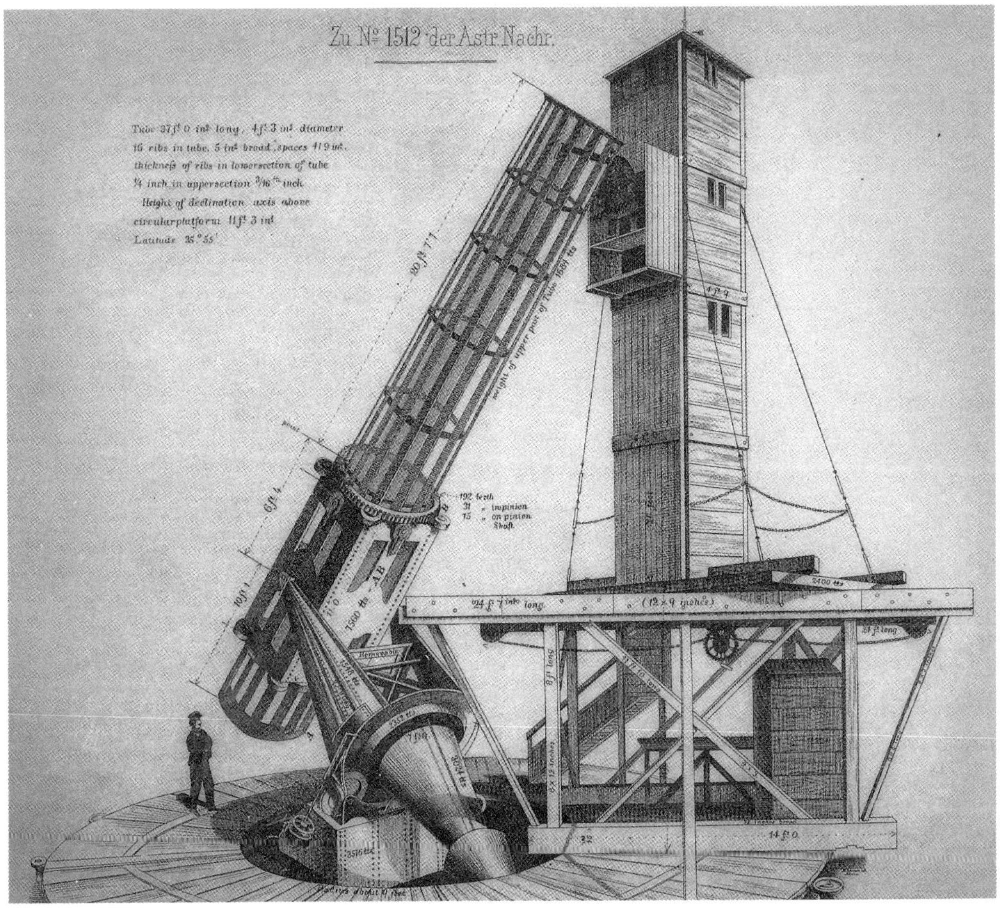

Fig. 2.5. The domeless telescope built by William Lassell (1799–1880).

position at the front end of the tube, with the help of an observing tower, which could be driven on rails around the central pier of the instrument. In the case of the Paris reflector, this involved a spiral staircase, and with Lassell's, a proper tower with an external lift and observation rooms on several floors. Lassell arranged the body of the telescope as a skeleton tube, in order to prevent certain possible thermal effects and to guarantee the optimum optical imaging quality. This was also the purpose of a complicated system of counterweights, which prevented deformation of the mirror through an accurately adjusted compensation system. The technical solution chosen for this purpose made it necessary to rotate the telescope tube about its main axis, according to the direction of observation. This was carried out on roller bearings at the load-bearing centre of the telescope.

A reflector with an exceptional design was built by the firm of Grubb in Dublin for the observatory in Melbourne. This instrument, with an aperture of 1.22 m (48″) was designed specifically to continue the programmes carried out between 1834 and 1838 in the southern hemisphere by John Herschel with one of his father's instruments, and also so that the astronomy of the northern and southern hemispheres could be pursued in a relatively balanced way. The project was prepared by a committee, which included, amongst others, experienced observers like John Herschel, Lord Rosse, Lassell and the Astronomer Royal, G. B. Airy.

Large twentieth-century reflecting telescopes

The decisive factor in the major rise in prominence of astrophysics that occurred after the turn of the century was the commissioning of large reflecting telescopes, most of which are still in use today. New questions concerning the nature and history of celestial objects, including the universe as a whole, required measuring weaker and weaker radiation fluxes and therefore large apertures that could not be obtained with refractors – the largest of which, as mentioned earlier, was at the Yerkes Observatory and had an aperture of 1.02 m (40″). It was the large and technically well-constructed reflectors, together with photography as a new detection technique, that first enabled serious study of extragalactic space, and of the true structure of our own stellar system, and permitted detailed spectroscopic examination of galactic objects.

Earlier than the famous telescopes of the Mount Wilson

Observatory, which will be mentioned later, there was another reflector that marked the beginning of the modern era of observational astronomy. In 1895 a 91-cm (36″) telescope was installed at the Lick Observatory, which had been founded seven years before, and, after complete reconstruction, this became famous as the Crossley Reflector. The previous owner, Edward Crossley had presented it to the Lick Observatory.

Chronologically later is the 1.82 m (72″) reflector built in 1918 for the newly founded Dominion Astrophysical Observatory near Victoria, British Colombia, in Canada. These telescopes are still in operation today and have provided a wealth of valuable observational material. The Crossley Reflector's contribution has been primarily in the form of long-exposure photographs of bright gaseous nebulae and galaxies, and in spectrographic measurements of radial velocities; and that of the Victoria reflector through exceptional spectroscopic plates of celestial objects. This type of material forms the essential observational basis for understanding the physics of such objects. The rapidly increasing amount of data obtained with these telescopes produced a correspondingly great increase in astronomical knowledge during that time. These telescopes became so significant because they were built at a time when new techniques in the construction of telescopes became available, and also when great advances had been made in photographic techniques.

No name is so closely connected with advances in the construction of telescopes as that of George Ellery Hale. Apart from special instruments for solar research, the four telescopes most famous today are fundamentally the results of his own work. With these telescopes, he quite consciously – and in a way quite characteristic of the turn of the century – followed the transition from refractors to reflectors.

This process began with the construction of the 40-inch (1.02-m) refractor for the University of Chicago's Yerkes Observatory (which is named after the donor). The optics came from the workshop of Alvan Clark and Sons, and the completed telescope, commissioned in 1897, surpassed all subsequent telescopes of this type. Hale recognized that a technical limit had been reached and realized that the development of larger optics could only be met by the reflecting-type of telescope. He accepted the necessity and therefore directed his organizational talents and his exceptional abilities as a designer towards the construction of a '60-incher' (1.5-m) telescope. It was designed for the Mount Wilson Observatory in southern California, which Hale had been able to found at the beginning of the century with financial assistance from the Carnegie Institution. The new institution was actually meant to concentrate on solar research in particular, and therefore the first instruments to be installed were a horizontal telescope, followed by a tower telescope. Hale was, however, able to justify his ambitious plans for such a large reflecting telescope with the fact that the examination of the stars would also make an important contribution towards the understanding of processes in the sun. This telescope was completed in 1908, and just nine years later, in 1917, Hale was able to commission a research instrument at Mount Wilson Observatory that was larger than any previous instrument. This was the 100-inch (2.5-m) Hooker telescope.

Hale's irresistible drive for progress is shown by the fact that he did not take on his projects one after the other: instead they overlapped chronologically. For example, he purchased the glass block for a 60″ mirror from France a year before the completion of the Yerkes Refractor, and in 1906, two years before the '60-inch' equipped with this mirror began operation, he turned to the businessman, J. D. Hooker in order to obtain the finance for a larger instrument. The mirror blank intended for this telescope was to have a diameter of 100″ (2.58 m) with an edge thickness of about 13″ (approx. 32.5 cm) on completion, and arrived on the day on which its smaller predecessor was finished. It became the main part of the Hooker telescope.

In the hands of some extremely able astronomers the two large telescopes at Mount Wilson Observatory have played a considerable part in the foundation of modern astrophysical knowledge. Striking evidence of this are the findings that were obtained by Edwin Hubble, Martin Humason and Walter Baade about extragalactic stellar systems. It was with regret, not solely based on nostalgia that the experts had to learn that the 100″ Hooker telescope was to be taken out of operation in the summer of 1985 for financial reasons. This probably means that the days of the '60-inch' are also numbered.

The breakthrough to telescope construction on a new scale was led by Hale, who proposed a project for a 200″ (5-m) telescope in 1928. With double the aperture and therefore four times the light-gathering area when compared with the Hooker telescope, it was to set the standards for a long time. Construction and operation of this giant telescope were planned in close conjunction with Mount Wilson Observatory, but the site chosen was Mount Palomar, approximately 150 km to the south-east. The deterioration in observing conditions that were already being encountered at Mount Wilson towards the end of the 1920s were taken into account in reaching this decision. The reason for this change was the increase in brightness of the night sky, caused by the rapidly developing city of Los Angeles and its surrounding towns.

Completion of the 200″ telescope brought about a great technical advance in astronomical instrumentation. This was not caused solely by the large aperture, but also by the inclusion of every possible technical improvement in the design and provision for fitting the auxiliary instrumentation required to investigate the incoming starlight. The new developments required the largest possible field, easy use of the primary and Cassegrain foci, availability of the Coudé focus in an air-conditioned room, and the ability to change rapidly between the different optical systems that were available.

This was a stage beyond the Hooker telescope, where no ideal solutions had yet been found for the points just mentioned. In particular, a new type of mounting was required that would allow observation of the North Celestial Pole and its immediate surroundings. In the case of the Hooker telescope – which was limited by its English-yoke-type mounting – the polar region, with a radius of 26 degrees, thus amounting to 10% of the northern hemisphere, remained inaccessible. The choice of material for the mirror and its design involved new principles when compared with the mirrors in earlier telescopes. These concerned improved thermal characteristics and a reduction in the weight of the glass blank. The 100″ Hooker mirror had been made from ordinary commercial glass and its optical quality was therefore strongly dependent on the extent to which it had reached equilibrium with the ambient temperature. This became apparent – much to Hale's dismay – during the first trials of the telescope, when the planet Jupiter appeared in the form of several, partially overlapping, discs. Fortunately, this was caused by a temporary distortion of the optical surface whilst it adjusted to night-time temperatures. Only after several hours, when equilibrium temperature was reached, did the mirror give its true optical performance. On that occasion, the situation was particularly bad, because it had been necessary to leave the slit in the dome open during the last stages of construction, and this had caused considerable heating of the instrument. But similar problems were undoubtedly bound to arise with the far greater thermal inertia of the 200″ mirror. It was therefore necessary to switch to a material with a lower coefficient of expansion. The use of quartz glass would have been technically feasible at the time – at least, experiments that had been carried out were promising – but this would have meant unbearably high costs. It was therefore decided to use *Pyrex* borosilicate glass, which, with an expansion coefficient of 2.5×10^{-6} per degree C was only one-third as sensitive to temperature change as ordinary glass.

The necessary reduction in weight was obtained without reducing the mirror's stability by using a set of ribs on the back of the mirror. For this purpose, cores were set into the mirror mould before casting. The manufacturer, Corning Glass Works in New York State, was also breaking new technological ground. There were some failures but, by the end of 1934, the second casting of the glass blank was successful. Compared with the 40 tonnes that would have been expected with the normal procedure, the colossus had a mass of only 20 tonnes. At the beginning of 1936, after its lengthy cooling phase, and after a complicated, but triumphal, journey right across the USA, the mirror blank arrived in the observatory's optical workshop in Pasadena, California, where further work was to be carried out. Here, the process of grinding and polishing went on for 15 years, albeit interrupted by the Second World War. It was the beginning of 1950 before the deviations of the mirror's surface were within the desired fractions of a wavelength of light of the required figure. The optical performance was so good that, at the primary focus, 80% of the light from a point source was concentrated into a disc 50 µm in diameter. It was clear from the very beginning that a mirror of this size would need a very carefully constructed cell in which special support systems would counteract the deformation that its own weight would cause with changes in position. This caused problems at first and the delay in completion can be partly attributed to the new design of mirror cell and support system.

So that the telescope might be flexible in its observing programmes, and in the use of auxiliary instrumentation, Cassegrain and Coudé foci were provided for spectrographic work; the primary focus being available for direct photography. A primary focal length of 16.75 m was chosen, relatively fast in comparison with the aperture, thus continuing a historical development towards simplification of technical problems and saving costs by requiring a correspondingly smaller dome. Starting with Herschel's large telescope of 1789, and leading up to the 100″ telescope on Mount Wilson, this development had introduced increasingly larger ratios between aperture and focal length. This, however, also had the disadvantage that coma, an optical aberration, became more noticeable, and eventually unavoidably limited the effective field of view. In the case of the 200″ telescope, with an aperture focal length ratio of 1 : 3.3, the primary field with relatively sharp images is only approximately two arc-minutes across. At Hale's suggestion, F. E. Ross had already designed a two-lens corrector system for the 100″ reflector, and this is put into the light-path just in front of the actual focus, considerably increasing the effective field. Ross also developed corresponding correctors for the large telescope. These extended its effective field to some 20 minutes of arc.

The possibility of producing aplanatic – i.e. coma-free – mirror systems was first demonstrated by K. Schwarzschild in 1905, although his systems had the disadvantage of long tube lengths and strong vignetting by the secondary mirror, which was situated in the light-path. Moreover, they were technically very difficult to make, because of the steep curvature on both mirrors.

Another solution, initially purely theoretical, was devised by Henri Chrétien and implemented by G. W. Ritchey on a 1-m reflector for the US Naval Observatory. In this system, the optical aberrations are also minimized by the use of complicated aspheric surfaces on the primary and secondary mirrors.

Hale had already worked with Ritchey during his time at the Yerkes Observatory. One of the smaller reflecting telescopes produced at that time had indeed contributed considerably towards Hale's conviction that reflectors were highly powerful. Ritchey later supervised the grinding and polishing work on both the 60″ and 100″ mirrors.

The most revolutionary solution to the problem of obtaining aplanatic mirror systems was the design developed by Bernhard Schmidt and first built in 1930, and which combined a wide field with outstanding imaging properties. The development of this system was to challenge refracting systems, which, until then, had been practically the only designs suitable for photographing wide fields of sky.

The 200″ mirror was naturally coated using the metal-evaporation technique that had been used since the beginning of the 1930s for astronomical purposes.

The physicists, P. Pringsheim and R. Pohl had discovered in 1912 that metals vaporized under a vacuum formed very good reflecting films on glass. After subsequent development of the process in the USA, by 1932 the best results were being obtained by the use of evaporated aluminium.

Such layers have a thickness of several tenths of a micrometre and, with appropriate care, are of outstanding homogeneity. The finest of the marks left behind from polishing are covered, reducing the amount of scattered light. The great advantage of evaporated aluminium films, however, lies in their high reflectivity, which reaches well into the ultraviolet region, and in their considerable chemical durability. This is in contrast to the silver layers precipitated from dissolved silver salts, and has ensured their widespread use. As a protection against mechanical damage, another thin layer of quartz is normally evaporated on top of the aluminium. For special purposes, dielectric coatings have nowadays become important but where, however, compromises have to be reached between their almost ideal reflective properties and the limitations in the effective wavelength range.

In view of the considerable mass of a 200-inch telescope, its special optical system, and the fact that a particularly stable mounting was required for the photographic programmes intended for the telescope, great effort was put into the design of the mounting and of the

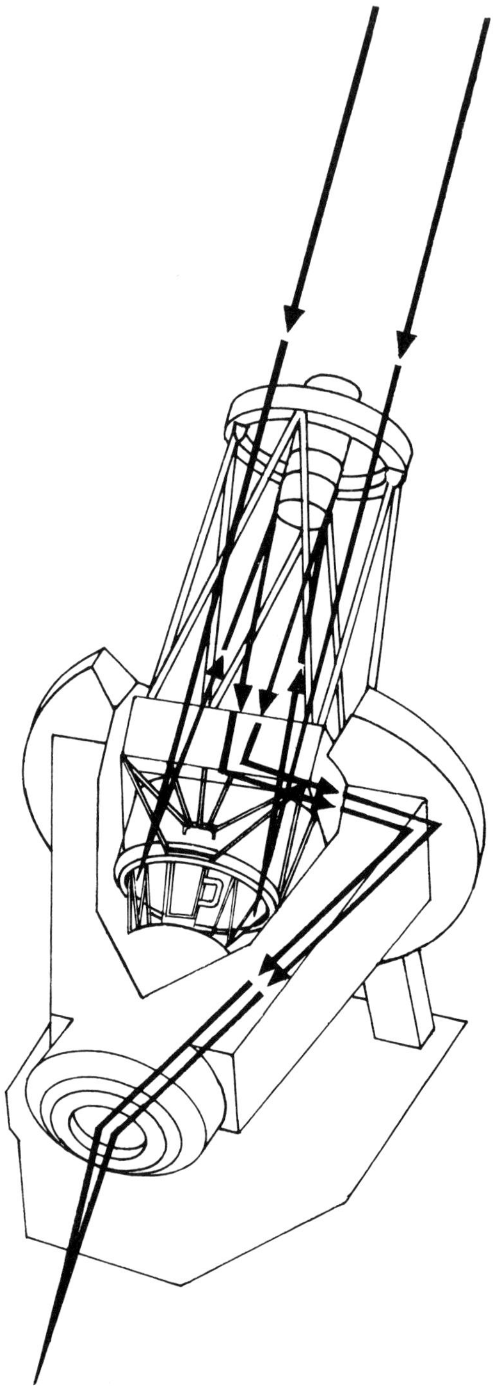

Fig. 2.6. The horse-shoe mounting of the 5-m Hale telescope, showing the optical path to the Coudé focus.

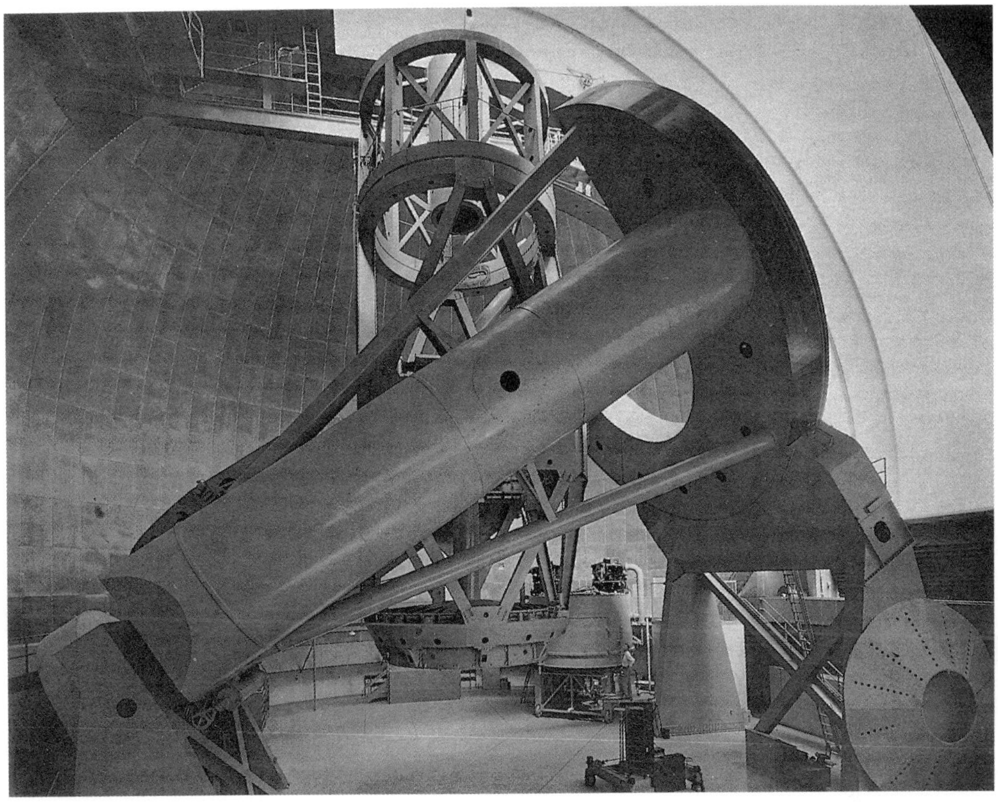

Fig. 2.7. The 5-m Hale telescope on Mount Palomar in California.

telescope tube. For the tube, M. Serrurier developed a special structural truss, which does not prevent mechanical bending, but which does ensure that both ends of the tube deflect by the same amount and without mutual distortion at different zenith distances. Thanks to this, under ideal conditions with the primary mirror at one end of the tube and either a Ross-corrector or a secondary mirror at the other, the optical system remains in constant alignment. This solution became the model for practically all, subsequent, large telescopes. This was also true for the horse-shoe yoke mounting which has become famous (Fig. 2.6). This ensured symmetrical suspension of the 140-tonne telescope tube, easy access for mounting equipment, and enabled the North Celestial Pole to be observed. Among other new techniques also explored here was the use of oil-bearing pads, which were fitted on both ends of the axis, and which meant that rotation of the whole 530-tonne mass was virtually frictionless. The careful design of the 5-m telescope (Fig. 2.7), borne out by its continuous success, has contributed a great deal towards understanding the mechanical processes involved in mountings, drive systems, mirror-support systems and other fundamental functions. This telescope has also pointed the way to later developments. Of course, it did establish some principles of construction that persisted, even when the state of technology had advanced in favour of other solutions.

Innovations in telescope design since 1970

In the case of optics developments moved away from the use of Cassegrain systems with correctors and towards the Ritchey–Chrétien form. Here the coma-free area is obtained for a field that is typically about 0.5° in angular diameter, using primary and secondary mirrors that have optical surfaces with curvatures expressed by higher-order mathematical functions. Correction systems are, however, necessary, in order to compensate for the spherical aberration that the non-parabolic primary mirror, with its large ratio of aperture to focal length produces at prime focus. In the meantime, however, the techniques for producing appropriate two- or three-lens correcting systems have become highly developed.

The employment of glass–ceramic materials has been crucial for the production and use of modern telescope mirrors. These are materials where, as a result of the production process, crystalline structures form within the otherwise amorphous mass of glass. The two components show different thermal behaviour, so a coefficient of expansion of zero – i.e. complete thermal stability – can be obtained by appropriate control of the production process. Trade names for these materials are Cer-Vit, Sital or Zerodur. Their advantages are revealed in astronomical observation because now the optical quality of the surfaces is practically independent of the mirror's degree of adjustment to the surrounding temperatures. The advantages have also been demonstrated in the production of astronomical mirrors, because polishing can be carried out more quickly and the resulting surface can be optically tested more frequently without the lengthy period previously required for the component to reach equilibrium.

As it is the heart of an optical telescope, the mirror's size and mass have a direct effect on the expenditure required for the mounting. It is not surprising therefore that, with the aim of reducing costs, there is a tendency to use thinner mirrors, or ones of light construction that are hollow with internal strengthening. The conventional diameter/thickness ratio was 6 : 1 but a ratio of 16 : 1 was chosen for the British 3.8-m infrared telescope sited on Mauna Kea on Hawaii since 1979. In spite of this, excellent optical quality was obtained, not only in the far infrared region, thanks to a system of 80 compensation devices on the back of the mirror. In future, quite new materials are definitely in prospect to carry the reflecting layer. Success has been achieved, for example, with synthetic materials strengthened by carbon fibres. An experimental mirror of this type has been produced that had a diameter of 1.3 m, but a mass of only 15 kilograms.

One major push behind research in this field has, of course, come from space flight, where the reduction in mass always plays a key role. It is notable that metals have also come under discussion recently as materials that are suitable for mirrors. Development has therefore come full circle – albeit at a higher level – back to Newton's work with metal mirrors that he began in the second half of the seventeenth century. This basic material is of interest today primarily because of its favourable manufacturing times and costs, and because of its high thermal conductivity. As far as large telescopes of the future are concerned, experiments are being made with nickel-coated aluminium and with various steel alloys.

Lightweight construction is now becoming popular, not just for mirror discs, but also for mechanical components and will certainly be a characteristic of future equipment. Such methods have become possible in large part because of the continuing developments in computer technology, which allow information from sensors, or mathematical flexure models to be continuously monitored, thus guaranteeing constant optimum alignment and adjustment of the optical system via suitable control elements. A special demand on sensor and computer technology is set by imminent forms of 'active optics' (Fig. 2.8). This aims at compensating for the deformation that occurs during operation by appropriately distributed pressure on the back of a relatively thin, primary mirror. Deformation is caused by varying gravitational effect when the telescope is in different positions, by thermal effects and, in certain conditions, also by wind. In practice, the images of stars produced by the telescope are subjected to high-speed analysis, enabling the appropriate corrections for the active stabilizing systems in the mirror cell to be calculated. Practical experience on a large telescope with a mirror diameter of 3.5 metres is being gained at present at the European Southern Observatory at La Silla in the Chilean Andes. Here, the 'New Technology Telescope' has been operational since early 1989, in which nearly 200 individual compensation elements will correct the shape of the mirror surface up to five times per second.

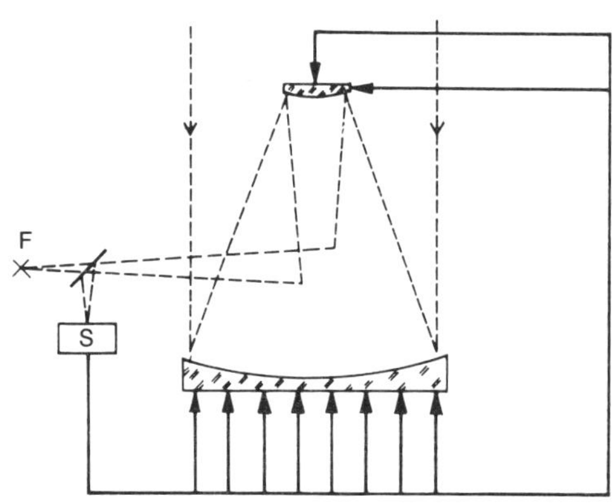

Fig. 2.8. A schematic representation of how active optics work. Part of the light entering the telescope is deflected before it reaches focus (F) and is directed towards a sensor (S). This produces controlled deformations (solid arrows) in the optical elements of the telescope (after R. N. Wilson, 1986).

'Adaptive optics' goes one step further in image optimization. In his book, *Opticks*, published in 1730, Isaac Newton wrote:

'If the Theory of making Telescopes could at length be fully brought into Practice, yet there would be certain Bounds beyond which Telescopes could not perform. For the Air, through which we look upon the Stars, is in a perpetual Tremor; as may be seen ... by the twinkling of the fix'd Stars. But these Stars do not twinkle when viewed through Telescopes which have large apertures. For the Rays of Light which pass through divers parts of the aperture, tremble each of them apart, and by means of their various and sometimes contrary Tremors, fall at one and the same time upon different points in the bottom of the Eye, and their trembling Motions are too quick and confused to be perceived severally. And all these illuminated Points constitute one broad lucid Point ... and thereby cause the star to appear broader than it is, and without trembling of the whole.'

Adaptive optics should serve to counteract this disturbance of stellar images by the Earth's atmosphere that was so clearly recognized and described by Newton. In a way similar to the process used in active optics, an image produced by the telescope will be analysed by suitable test methods. This time, however, it is important to obtain information about the distortion of the incident wavefront caused by fast-moving, atmospheric turbulent cells with varying refractive indices. An auxiliary mirror placed in the telescope's light-path has to be deformed in real time, so that the overall system, consisting of atmosphere and telescope, maintains a perfect optical image, which ideally, is limited only by diffraction at the entrance aperture. With larger future telescopes in particular, this will make it possible to increase the image quality to such an extent that, for the first time, the measurements will fully justify the high investment costs. The computing power required for the rapid image-analysis and subsequent correction of the phase of the incident waveform is extraordinarily high, because the control frequencies are around 200 Hz with a very large number of individual points. Pilot systems have already given promising results, and the possibility of employing adaptive optics is planned for all large, future projects. As the individual turbulent cells have dimensions of approximately 20 cm, and occur in the atmosphere up to altitudes of several kilometres, the wavefronts differ for stars that are only a few seconds of arc apart. Initially, therefore, adaptive optics can only optimize imaging of individual objects.

Since about 1970, computers have come to be used for controlling the positioning of telescopes. Electronics have relieved the observer of calculating the necessary coordinates, of the tricky manoeuvring involved in getting the telescope into position, of supervising telescope functions, of monitoring the position of the observing slit, and of similar tasks. A whole field of automated control has since been developed.

With a modern instrument, the control of elements, position encoders and a whole hierarchy of computers are coordinated and integrated into a highly efficient system. In addition it can control the instrumentation on the telescope and capture the data, as long as the latter are suitable for digital processing. People no longer need to control the activity in the dome, and can concentrate on the incoming data and their initial interpretation. They no longer have to work on special platforms or from the primary- or Cassegrain-focus positions of large telescopes, but from comfortable control rooms. This can be taken so far that, just as with the control of astronomical equipment mounted on satellites, a ground-based telescope can be controlled through the ordinary communications networks from an institute based in another continent. Some examples of this kind of set-up are already in existence. British astronomers, for example, have introduced a direct data-link to their infrared telescope on Hawaii, and recently the 2.2-m telescope of the European Southern Observatory in Chile has been controlled from the centre in Garching near Munich, Germany. Not only were images of finder fields obtained by TV cameras relayed to monitors at the control centre, but the actual observational data and digitized images were also transmitted to the computer memory banks. The amount of time necessary to transmit a stellar spectrum amounted to about two minutes. It proved to be convenient to keep the night assistants at the telescope during its operation and not, as was originally planned, to control all the telescope's functions over such great distances.

The Hale telescope, and several large telescopes of the 4-m class, are fixed on a horse-shoe-type mounting, despite the relatively large amount of material required, because their symmetrical form gives them considerable mechanical advantages. During the construction of the 6-m telescope at the Special Observatory of the Soviet Academy of Sciences – at present the largest optical astronomical instrument – the whole question of the mounting was reconsidered. It was precisely for reasons of symmetry that the classical equatorial mounting with its inclined polar axis was forgotten in favour of an altazimuth mounting with one vertical, and one horizontal axis, to support the massive tube of the telescope. The earlier, far more massive, radio telescopes served as examples. Because, with this type of mounting, one axis of rotation is always parallel to the direction of gravity, the bending moments affecting the tube are only dependent on the inclination of the tube and always act in the same cross-section of the tube. The vertical rotation axis remains free of flexure. The motion of the telescope, corresponding to the apparent daily rotation of the sky, becomes more complicated, and, in addition, a rotation of the field of view takes place, but these can be relatively simply controlled by computer software. With the mounting, however, the region immediately surrounding the zenith remains inaccessible, because there the azimuth of the telescope would have to be altered extremely rapidly accompanied by fast rotation of the field of view. This does not impose serious limitations on the use of such telescopes, because any observational programme can always be appropriately modified.

The 6-m telescope (Fig. 2.9) was set up at the beginning of the 1970s, on the northern edge of the Caucasus near the village of Zelenchukskaya, and began operation in 1976. The mirror, produced from a 60-tonne block of Pyrex-like borosilicate glass, can be used at prime focus with a focal length of 24 metres. Alternatively, the light-path may be extended and reflected through the hollow altitude axis to observing platforms carried by the main load-bearing fork. From an optical point of view, the whole system is conventional, but this type of mounting became the pattern for telescopes subsequently constructed. The altazimuth type of mount has already been used in smaller telescopes, for example, in the 2.3-m Advanced Technology Telescope at Mount Stromlo and Siding Spring Observatories in Australia. Efforts began ten years ago at that institute to find a cost-effective instrument for research and training. In 1984 the ATT with its advanced technology became available, demonstrating the advantages of a simplified mount, a thin mirror (22-cm edge thickness) and complete computer-control. Without exception, all projects for future large telescopes are making provision for an altazimuth mount.

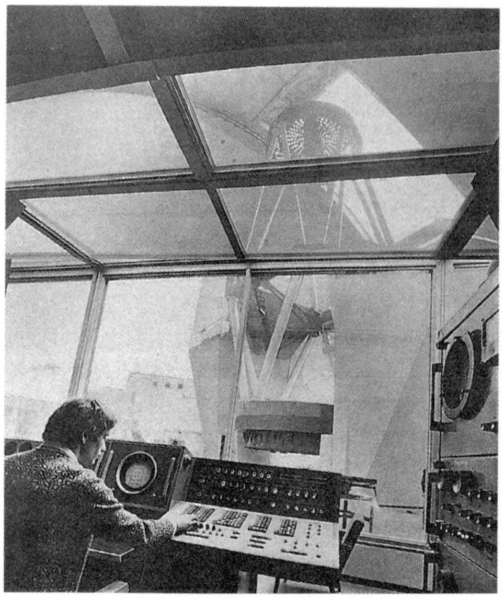

Fig. 2.9. View of the 6-m Soviet telescope near Zelenchukskaya.

An altazimuth mount also has a considerable advantage as regards the dome that protects the telescope from the influences of the weather. Because of the simplified motion of the telescope, this can be built to enclose the tube and mounting closely. An example is the cube-shaped building that protects the 'Multi-Mirror Telescope' on Mount Hopkins in Arizona, and which rotates with the telescope in azimuth. The effects of an astronomical dome are under much discussion at present. Its task is to protect the expensive instrument from the rigours of the weather, from the hot solar radiation during the day and from wind pressure at night. At the same time, however, it prevents complete temperature equilibrium between the mirrors and the surrounding layers of air from being reached and the danger arises that it may cause additional turbulence immediately around the telescope, thus interfering with the image quality that can be obtained. There is therefore a tendency towards the construction of free-standing, dome-less telescopes. The 'made-to-measure', protective housing of an altazimuth mounting might be a good compromise between the two extremes.

Telescopes outside the Earth's atmosphere

The future of observational astronomy probably lies equally with telescopes in space and a new generation of ground-based telescopes, whose light-collecting surfaces will have diameters of over ten metres. The construction and operation costs of telescopes in space are so high that they will only be used for research programmes that cannot be carried out from the ground because of the limitations set by the atmosphere. It has been estimated that the Hubble Space Telescope, with a mirror diameter of 2.4 m, operating at an altitude of 600 km, will cost ten times as much as a ground-based 10-m telescopes. Measurements that are particularly suitable for the astronomical observatories operating in orbit are those in the ultraviolet and infrared spectral regions, and at present, those programmes that require high angular resolution. With the former type of observation (in IR or UV), the limitations of ground-based work are set by atmospheric transmission, which is strongly dependent on wavelength. With the latter, even the highest layers of the atmosphere are subject to turbulence that prevents the theoretical optical resolution being attained, and requires costly procedures such as adaptive optics if it is to be overcome.

After the launching of the very first artificial Earth satellite, 'Sputnik' on 4 October 1957, astronomical investigations were very soon included in the research programmes of both the major space powers. Some of these soon concerned objects outside the Solar System. As far as their aims and methods of implementation are concerned, such programmes are closer to classical methods of astronomical research than spaceprobe missions to investigate planetary geology. Nevertheless, both from a scientific and a technological point of view, they all obviously belong to a separate, and already extensive, chapter in the recent history of science.

Instrumentation had to be produced that would function under the conditions in space and that would reliably provide observational data on celestial objects, either following pre-set programmes or under command from control centres on Earth. Astronomical satellites designed for such investigations consist of an imaging telescope, measuring instruments and detectors, telemetry units, power supplies and, in some cases, attitude-control systems.

Thanks to these developments astronomical knowledge has progressed in leaps and bounds since 1957, and we have gained new depths of understanding about processes occurring in the universe. To a large extent this can be attributed to satellites such as 'Copernicus' (launched in 1972 by the USA), 'International Ultraviolet Explorer' (1978, USA and Western European countries) and 'Astron' (1983, USSR) covering the ultraviolet region; and at infrared wavelengths between 12 and 100 μm, to the highly successful 'Infrared Astronomy Satellite' – IRAS – (1983, The Netherlands, USA, Great Britain). It is also possible to focus radiation at X-ray wavelengths, although with such short wavelengths, one has to use approximately cylindrical mirrors the inner surfaces of which bring grazing-incidence radiation to a focus. The Einstein satellite, launched by the USA in 1978 was the first to provide images of celestial objects at wavelengths below five nanometres. Other important steps in this field were marked by the Qvant astronomical module which was coupled to the Soviet space station 'Mir' in 1987 and especially the X-ray satellite ROSAT launched in summer 1990.

The whole international scientific community is waiting with great suspense for the launch of the Hubble Space Telescope as for no other, earlier astronomical instrument. The launch of the largest space telescope was, after several postponements, early in 1990. The flood of data that it will provide will cover all types of celestial objects and will have an effect on the entire range of astronomical sub-disciplines. It will form a new basis for answering a variety of questions ranging from cosmology through to the direct detection of individual planets around other stars.

The space telescope named after Edwin Hubble started with plans in the USA for a 'Large Space Telescope'. The design had been more or less finalized but then in 1972 the original three-metre diameter had to be reduced to 2.4 m because of financial reasons. The new plan was scheduled to be completed in 1983 but several postponements were required because of difficulties that several contractors encountered. The eventual launch depended on the readiness of the Space Shuttle that was to carry the telescope into orbit.

The life of this satellite telescope is planned to be at least 15 years but interim maintenance might be carried out by specialists using the Space Shuttle. The instrumentation, though, is very versatile and is designed for long use. The choice of the measuring instruments on board is based on a quite specific, but broadly based, concept of the work to be carried out. The Space Telescope Science Institute in Baltimore was specially founded some years ago, and is responsible for the operation of the Space Telescope. There is a contact point for Western Europe in Germany. The various detectors in no fewer than eight instruments, covering a range of wavelengths from the middle ultraviolet (115 nm) to the near infrared (1.1 μm), convert radiation into digital signals that are sent to the Institute via satellite radio links. Table 2.1 gives some details of the instruments used on the Hubble Space Telescope, and indicates how advances in astronomical technology have not been confined to just the telescopes themselves. Improvements in auxiliary instrumentation have also played an important part.

In order to reduce weight, the mirror for the Space Telescope, which was finished by the end of 1981, was made as a sandwich construction, consisting of two thin glass plates rigidly supported by an intermediate structural honeycomb. The material chosen naturally had a very low thermal coefficient of expansion. The type of construction meant that the mass was reduced to about 20% of that of a solid mirror of the same size. The surface of the mirror was planned to have a standard deviation of 10^{-5} mm from the ideal hyper-

Table 2.1. *Instruments on board the Hubble Space Telescope*

	Aperture ratio	Field size or slit width	Resolution	Spectral range in nm	Limiting magnitude
Wide-field/	*f*/12.8	2.7′ × 2.7′	0.1″	115 to 1100	9.5 to 28
Planetary camera	*f*/30	1.2′ × 1.2′	0.04″	115 to 1100	8.5 to 28
	Tasks: Direct exposures through filters, polarimetry				
	Detector: Mosaic of four CCD matrices each having 800 × 800 pixels				
Faint object camera	*f*/96 or	11″ × 11″	0.02″	120 to 600	21 to 28
	f/48	22″ × 22″	0.04″	120 to 600	21 to 28
	Tasks: Direct exposures through filters, polarimetry, spectroscopy of two-dimensional objects. Able to mask bright nearby stars (using the coronograph principle)				
	Detector: Image intensifier followed by TV tube				
Faint object		0.1″ to 4.3″	0.3 nm or	115 to 700	19 to 22
spectrograph			3 nm	115 to 700	22 to 26
	Tasks: Medium resolution spectroscopy, polarimetry				
	Spectral analysis: Choice between six different diffraction gratings and one prism				
	Detector: Two Digicon detectors, each having 512 silicon diodes, and covering half of the range of wavelengths				
High resolution		0.25″ to 2″	0.003 nm or	110 to 320	brighter than 11
spectrograph			0.015 nm or	110 to 320	brighter than 14
			0.15 nm	110 to 170	brighter than 17
	Tasks: Spectroscopy with high spectral and time resolution (50 ms)				
	Spectral analysis: Choice between five different diffraction gratings and one echelle grating				
	Detector: Two Digicon detectors each with 512 silicon diodes				
High speed		0.4″ to 10″	16 μs	120 to 800	brighter than 24
photometer	Tasks: Photometry in numerous wavelength regions, measurement of the change of brightness when stars are occulted by the Moon and planets				
	Detector: Dependent on wavelength and specific applications, image-dissectors and one photomultiplier				
Three identical	Non-rectangular field,		0.003″	460 to 700	4 to 20
fine guidance	69 square minutes of				
systems	arc in area				
	Tasks: Fine alignment of the telescope, astrometric measurements				
	Detectors: Each has four photomultipliers				

boloid. An exceptionally high percentage of radiation was expected to concentrate in a disc, much smaller than can be attained with any ground-based telescope where the formation of the images is limited by atmospheric turbulence. In reality HST's optical system suffers from a severe amount of spherical aberration. This has now been tracked down to a fatal error made during the testing phase of the main mirror. Only 15% of the light is imaged within a core 0.2 seconds of arc diameter, the rest is spread over an area 3 seconds of arc across. As was proved during its first months of operation HST is nevertheless capable of doing good science. At least for bright point sources computer processing can improve the resolution of images far above any previous results. Modification of the optical system during maintenance can be expected for the end of 1993.

As usual the mirrors' reflecting layers are made of evaporated aluminium, but, in this case, it is overcoated with magnesium fluoride (MgF_2) for protection and also to increase the coefficient of reflection in the ultraviolet spectral region. Close to the short-wave working limit, a reflection coefficient of 75% has been attained; in the red spectral region this amounts to 89%. Pointing the telescope towards the objects that are on its observational programmes, is accomplished by sensors. Two of the three sensors lock onto the positions of fixed stars. The sensing systems are able to determine errors in the positioning of the telescope relative to the stars with an accuracy of three milliarcseconds, and are able to ensure that the alignment remains within seven milliarcseconds over an observational time of several hours. This is far more accurate than the performance that could be obtained from any ground-based telescope. The great accuracy of these sensors can only be used to the full if, over the whole sky, there is a sufficiently large number of accurately measured star positions available as reference points such that at least one falls within the small field of view of one sensor. In order to ensure this, a great deal of preparation was necessary, during which the plates of the two extensive atlases produced with Schmidt telescopes were scanned. Based on the Southern Sky Survey that had been made with the 1.2-m United Kingdom Schmidt at Siding Spring in Australia, and a special 'Quick V' survey made with the 48″ Big Schmidt at Mount Palomar, a computer catalogue was produced containing the positions of nearly 19 million stars, measured to an accuracy of 1.7 seconds of arc.

The accuracy of the position sensors is so high that it is worthwhile using the sensor that is not used for the positioning of the telescopes for astrometric programmes. The errors are expected to be one-tenth of those of ground-based astrometry, which means that results should be obtained that are of fundamental importance for determining the distances and proper motions of stars.

The Hubble Space Telescope's instrumentation uses the same highly advanced detectors that are employed at modern, fully equipped ground-based observatories. Regardless of whether the equipment used is a large ground-based telescope or a satellite telescope, the methods in which it operates are determined by the problem being studied and by the physical and technical procedures available. Irrespective of the type of observational platform, similar streams of data are fed to appropriate computers and are analysed with partially, or completely, identical data- or image-processing program. The technical problems involved, the methods used and the instrumentation and processing, all mean that Earth-based astronomy and that carried out from orbit are, in fact, very closely related.

New ground-based telescope projects

The potential of the Hubble Space Telescope has, in fact, reinforced demands for new ground-based telescopes with larger reflecting surfaces. Projects of this type can be found at the top, or at least near the top, of different countries' lists of priorities for major investments in astronomical research. One of the reasons for this is the necessary extension of the range of spectroscopic work to include fainter sources of radiation. The spectra of the faintest galaxies identified and classified by the Hubble Space Telescope, for example, will only be observable by very large telescopes, and therefore, for the foreseeable future, only by ground-based instruments. The recognition that basic questions concerning the distance, motion, chemical composition, and physical conditions of galactic and extragalactic objects cannot be answered by imaging or photometry in integrated light alone, but only by the examination of spectral details, is, of course, not new. Telescopes built this century were therefore often designed for spectroscopic use. What is new is that people are no longer satisfied with exceptional optical imaging only near the optical axis, as in the classical Cassegrain system. There is also a strong tendency for spectroscopy to require a larger field of sharp stellar images, thus making better use of a large telescope by studying several objects at once. This has been a matter of course for a long time in photographic work, and, in the past, provided the main stimulus towards the development of the Schmidt system.

Apart from climatic and local atmospheric conditions, the natural brightness of the night sky and light-pollution caused by civilization have a very serious effect on the search for very faint objects. Careful choice of the correct geographical location is therefore of the utmost concern. The reduction of interference by background illumination also introduces technical demands that play a great part in the design of a new-generation instrument. These demands involve the need for the highest mechanical stability and, at the same time, reduced construction and costs, and the best quality imaging, taking both atmosphere and telescope optics into account.

Among the serious projects for new, large, future telescopes that have been under discussion since about 1970, the following techniques are of particular significance:

1. Production of a monolithic lightweight mirror. This is conceivable up to a diameter of about 8 m. With dimensions much above this, there would be insurmountable problems of transportation and in accommodating the mirror in a vacuum chamber for coating it with the highly reflecting surface.
2. Producing a large mirror from a mosaic of individual sectors or hexagonal 'tiles'.
3. Splitting the optics into several individual systems. A further

sub-division involves placing these separate systems on a common mounting, or on individual mountings.

One installation of the last type has already been realized.

As a pilot development for multiple-mirror telescopes on a single mounting, the Multi-Mirror Telescope (Fig. 2.10) was constructed at the Whipple Observatory on Mount Hopkins in Arizona by the University of Arizona and the Smithsonian Institution's Astrophysical Observatory. The idea for such a telescope arose at the end of the 1960s, from an analysis of production costs versus telescope aperture. According to this study, a combination of several instruments of between 1.5 m and 2 m proved to be particularly cost-effective. Special circumstances prompted actual construction and, by the spring of 1979, the prototype was available for astronomical research. Six complete Cassegrain systems are 'connected in parallel' and share one mounting. The six mirrors are made of lightweight material, and each has a diameter of 1.8 metres. The accompanying secondary mirrors are held by a compact skeleton tube. Additional optical components bring the light-paths together to a single focus and superimpose the resulting individual images. A sophisticated control system is responsible for steering the secondary mirrors in three dimensions in order to hold the focus and fine alignment of the combined images stable over long periods of observation.

As far as the total light collecting surface is concerned, this instrument on Mount Hopkins is equivalent to a telescope with an aperture of 4.5 metres. The concept has proved to be of promise for future large telescopes. As a result, a project for a telescope consisting of four 7.5-mm mirrors has been proposed in the USA, and the telescope could begin operation towards the end of this century. Considering the price per square metre of light-collecting surface, this telescope compares very favourably with the Hubble Space Telescope. Apart from its great light-grasp for the examination of stellar and extended objects, a significant advantage would lie in its high angular resolution. This is a result of the large diameter of the overall system, provided that the radiation from the individual telescopes can be combined without phase distortion.

Also in preparation, but apparently at a far more advanced stage than the project just mentioned, is the 'Very Large Telescope' currently being developed by a large team for the European Southern Observatory (ESO). After taking several factors into consideration – such as flexibility during subsequent operation, previous experience, and the fact that the system was to be completed as soon as possible – it was decided to construct four telescopes alongside one another. Experience was obtained in particular from the construction of ESO's 'New Technology Telescope' which has already been mentioned. It is planned to incorporate equipment beneath the observing platforms that will be capable of bringing two or more of the individual bundles of light to a common focus. Possible methods of combining the images are: the use of direct optical imaging, of fibre optics, or indirectly by combining the electrical signals from the detectors. The design uses a mirror diameter of eight metres – i.e. it is pushing current technology to its

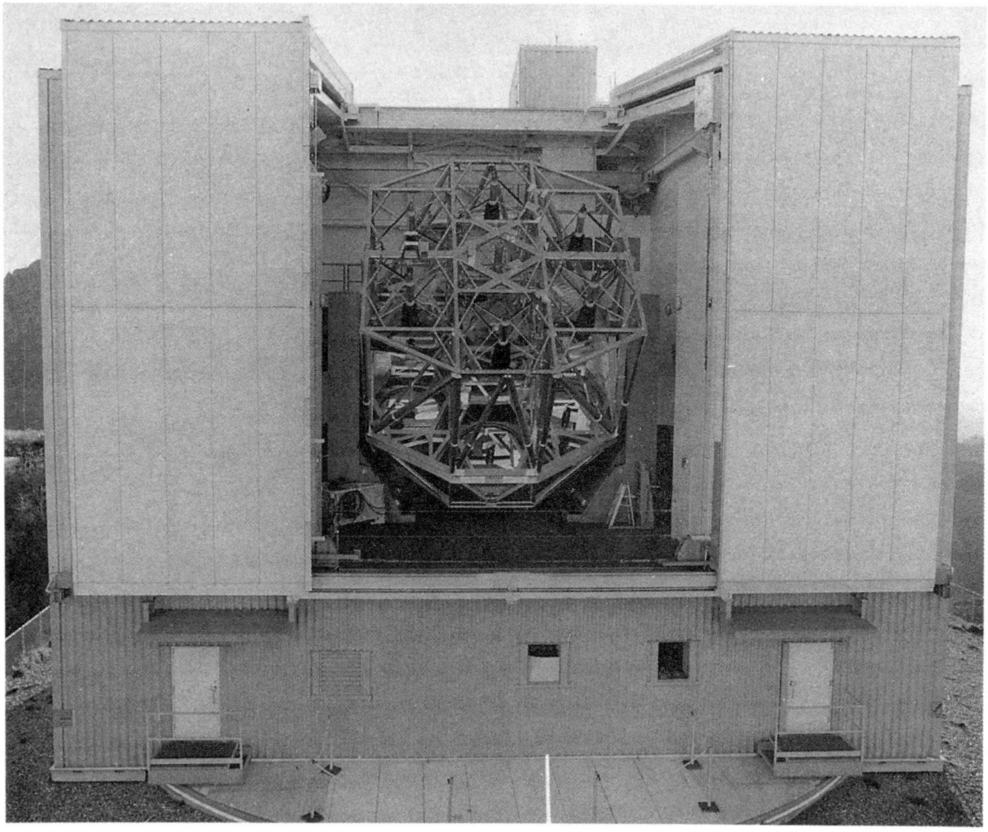

Fig. 2.10. The Multi-Mirror Telescope on Mt Hopkins in Arizona, which represents a new generation of astronomical telescopes.

limit. This would achieve the equivalent of a 16-m telescope. This corresponds roughly to the American project for a large multi-mirror telescope, so comparable results can be expected. ESO's design appears to be superior as far as interferometric work is concerned, because the distance between the two outer telescopes offers a baseline of 100 metres, and will therefore allow a higher angular resolution to be obtained. The mirrors are to have aperture ratios of *f*/2, and it is, of course, planned to have active correction of the shape of the mirror during operation. The telescopes are to be built without a dome.

Special aerodynamic studies are being carried out to determine the best protection from the wind, and it appears that, with adaptive optics, individual platforms around each telescope, and various means of directing and controlling the airflow, the large Venetian-blind-type windscreen originally proposed may no longer be required. If all proceeds according to schedule the first of the four telescopes should be available in 1993 and the whole system should be completed five years later.

There are two other future projects involving the type of instruments fitted with mosaic mirrors that was mentioned above.

In the Soviet Union, an examination is currently in progress of the scientific and technical aspects of building an astronomical telescope with a mirror surface of about 500 square metres, corresponding to a diameter of around 25 metres. So far a spherical main mirror has proved to be the most favourable solution, and this would be assembled from approximately 500, actively controlled, hexagonal sections. The gigantic dimensions of such a telescope become particularly clear when one considers that the secondary mirror alone, with a diameter of six metres, will be as large as the primary mirror of the largest contemporary telescope. The chosen design places the primary and secondary mirrors a distance of 55 m apart, and the total mass is estimated at about 2000 tonnes. With this enormous project in mind, scientists have been gaining experience of a segmented mirror with a diameter of 1.2 metres, and its control system, for the past ten years at the Crimean Astrophysical Observatory of the Soviet Academy of Sciences.

Two scientific institutions in the USA are planning joint construction of a 10-m telescope for the beginning of the 1990s. In this case, the primary mirror is to consist of 36 hexagonal parts each 1.8-metres in size. With appropriate active alignment, the single components should behave as a single large paraboloid. This will, however, cause considerable problems during figuring, because each part must have a different shape depending on its position in the overall surface. The site chosen for this project, the Keck Telescope, is Mauna Kea on Hawaii, already the home of several observatories. Mauna Kea is also envisaged as a telescope site by Japanese astronomers. They hope to install a telescope there and to begin operation in 1993. This is planned to have a monolithic 7.5-m mirror and is intended to work in both the optical and infrared regions.

The projects described in the last paragraphs may be somewhat vague, but it can be seen that promising developments are taking place in this area of astronomical observational technology. There would seem, therefore, to be no doubt that questions that are still unresolved, and which in some cases have only recently been raised, will, in future, be tackled with appropriate instruments and techniques.

3 The Schmidt telescope

In astronomy there are numerous, varied, observational objectives that have to be met. On the one hand, these may involve obtaining photometric or spectroscopic data about individual objects, and on the other, obtaining information about a number of celestial objects simultaneously. Whereas the former only require small fields of view, the latter entails faultless imaging over large fields.

It is easy to obtain coma-free fields with large angular diameters when using lens systems, but objective glasses are always subject to chromatic aberration. The aberration known as coma causes rays of light that are incident at an angle to the optical axis to form asymmetric, comet-like images. This causes great problems when it comes to evaluating observational material. Chromatic aberration means that radiation of different wavelengths (i.e. different colours) is refracted by different amounts. There is no chromatic aberration when mirrors are used, but mirrors only produce coma-free images for rays that enter parallel to the optical axis, which means that classic paraboloids have a very small field of view. Optical workers and astronomers have naturally struggled hard to remove the errors occurring in various optical systems. With reflecting telescopes, this means using suitable correctors so that even off-axis rays give coma-free images, and the resulting telescope has a large field of view.

The first man to carry out a thorough investigation of the theory of such designs was Karl Schwarzschild, 1873–1916 (Fig. 3.1). He spent a lot of time investigating the theory of aberrations of optical systems in general and, using this as a basis, analysed the performance of parabolic mirrors that were then in general use for astronomy. Once familiar with the theory of the optical aberrations of parabolic mirrors, Schwarzschild set himself the task of calculating reflecting systems that were free from both coma and spherical aberration. Spherical aberration means that rays incident at different distances from the central axis have different focal lengths.

In the best result that Karl

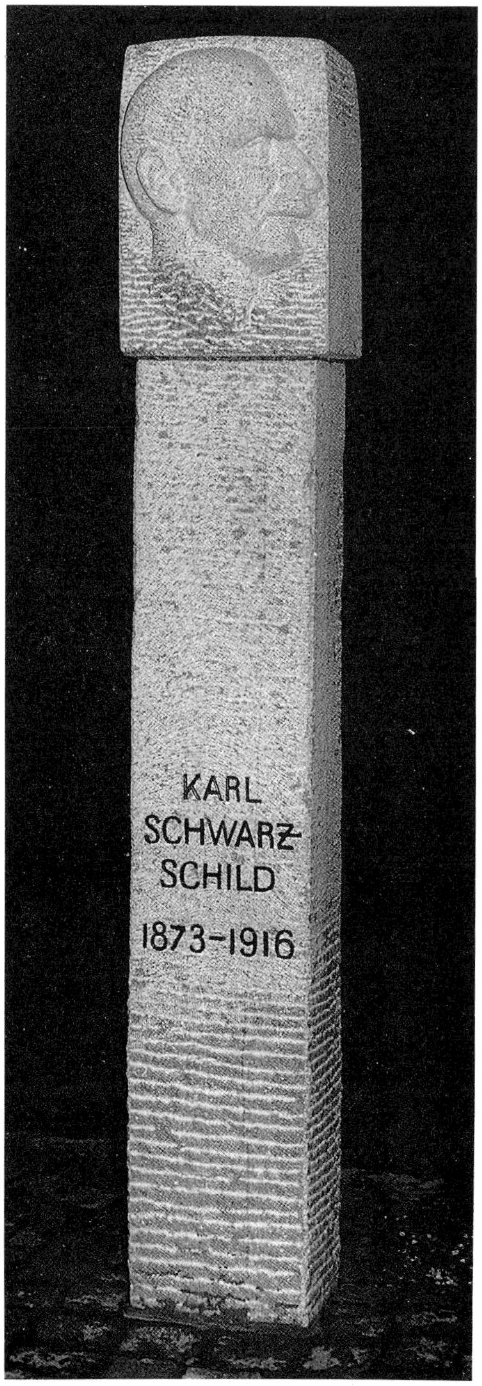

Fig. 3.1. Karl Schwarzschild was born on 9 October 1873 in Frankfurt/Main. At the age of 28, he became Director of the observatory at Göttingen and in 1909 that of the Astrophysical Observatory at Potsdam. Karl Schwarzschild carried out important work on photographic photometry, fundamental investigations into the theory of stellar atmospheres, the proper motion of stars, and stellar statistics. In addition, he was greatly involved with problems concerning astronomical optics. This memorial stands in the grounds of the Karl-Schwarzschild Observatory at Tautenburg. It was made by the Berlin sculptor Wehrmeister.

Schwarzschild achieved, the primary mirror was a hyperboloid, and the secondary mirror a concave ellipsoid, the minor axis of which coincided with the optical axis (Fig. 3.2). The system had a focal length of one metre and the primary mirror a diameter of 33 cm. So Schwarzschild had produced a very fast system with an aperture ratio of 1 : 3. With this system, the images of the stars over a field having a diameter of 3° were circular, and thus coma-free. The diameter of the images at the edge of the field was 16 seconds of arc across. Such large stellar images meant, however, that the radiation was distributed over a very large area of any detector – e.g. a photographic plate – which naturally entailed a loss of limiting magnitude.

If we assume for the moment that the best images permitted by atmospheric turbulence have a diameter of 2 seconds of arc, and that such image quality is attained by images at the centre of the field, then the images at the edge of the field with diameters of 16 seconds of arc involve a loss of approximately 4.5 magnitudes. Even if the increase in size of the images from the centre to the edge of the field can be determined systematically, and thus be expressed accurately in mathematical terms, the photographic reduction of such plates still presents considerable difficulties.

This disadvantage was certainly one reason why Schwarzschild's reflecting system found no astronomical applications. Another was that the secondary mirror had to be very large, and therefore introduced a large central obstruction. A system with a focal length of one metre and a primary mirror 33 cm in diameter requires a secondary mirror with a diameter of 16.7 cm. With normally incident light – i.e. rays that travel parallel to the optical axis – this obstructs 25% of the primary mirror. Another point that should be mentioned is that – at least in Schwarzschild's time – the production of hyperbolic and ellipsoidal surfaces on primary and secondary mirrors presented considerable technical difficulties. Despite the fact that the Schwarzschild system was never used in astronomy, it showed for the first time, that, in principle, the production of coma-free systems is feasible.

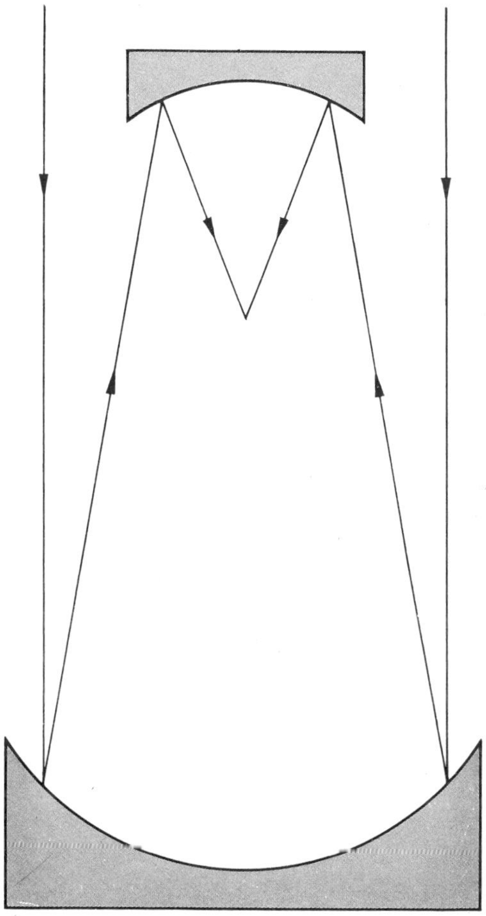

Fig. 3.2. The Schwarzschild system consists of two concave mirrors whose concave surfaces face each other. The radius of curvature of the primary mirror (a hyperboloid) is 3 m, that of the secondary (an ellipsoid) is 1.67 m. The focal length of the system is 1 m, so that a primary diameter of 33 cm gives a focal ratio of f/3.

A further means of removing the aberrations inherent with a parabolic reflector is by the insertion of special, correcting-lens systems. In this way, one can virtually overcome the coma, at least in principle, but then chromatic aberration, which always occurs in lens systems, has to be taken into account. Apart from Karl Schwarzschild's reflecting system just mentioned, and the use of lens-corrector systems, there have been several other attempts at producing reflecting systems with large aberration-free fields. All of the methods, however, were encumbered with considerable faults. Everyone who had investigated this problem had used the classic parabolic reflector as a starting point (Fig. 3.3).

Bernhard Schmidt's idea

Bernhard Schmidt started from a completely different basis. He chose a spherical mirror as the primary mirror. With a spherical mirror there is no particular optical axis, i.e. all bundles of rays, coming from any direction whatsoever, are completely equivalent. Images are generated on a spherically curved

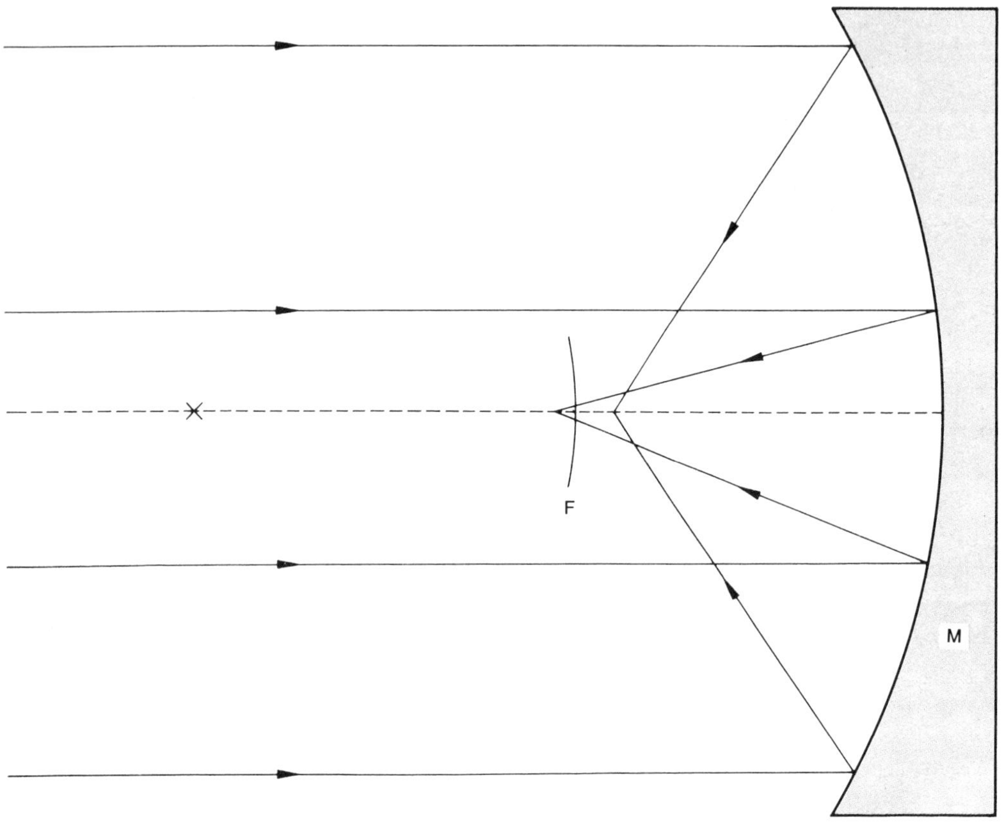

Fig. 3.3. The distance of the curved focal surface (F) amounts to half of the radius of curvature of a spherical mirror (M). As a result of spherical aberration, the central rays of a pencil of light come to a focus farther away from the surface of the mirror than the peripheral rays.

surface. The radius of curvature of the image surface amounts to half the radius of curvature of the reflecting surface. Moreover, the centres of curvature of the spherical mirror and of the image surface coincide. At this common centre of curvature, Bernhard Schmidt placed an aperture stop. When used with an aperture stop, a spherical mirror does not suffer from coma, so no compensation for this aberration is required. Although coma-free, images produced by a spherical mirror are not entirely free of distortion. Spherical mirrors are subject to spherical aberration, i.e. the central bundles of rays have a longer focal length than those at the edge of the mirror. Bernhard Schmidt's ingenious idea was to remove the spherical aberration by means of a special lens. This special lens has to increase the focal length of the outer rays and decrease that of the central rays. It therefore acts as a diverging lens for the outer zone and as a converging lens in the central area. Today this lens is known generally as a 'Schmidt plate' or a 'corrector plate' (Fig. 3.4).

Bernhard Schmidt's method of producing the corrector plate was just as inspired and simple as his original basic idea. He laid the plane-parallel plate that was to be worked, on a perfectly circular chamber with the same diameter as the plate. Bernhard Schmidt sealed the plate onto the edge of the chamber with oil or vaseline and then exhausted the air out of the chamber. This causes the plate to sag in the centre. The extent of this sag depends on both the strength of the glass plate and on the pressure difference to the ambient atmosphere. With the plate held in this state, Schmidt ground and polished a circular zone close to the edge, as shown in the sketch. When the air was allowed back into the chamber, the plate took on the required shape – it acted as a converging lens in the centre, and as a diverging lens in the outer zone. It thus increased the focal length of the outer rays and shortened those of the central rays after they were reflected by the spherical mirror.

This process that Schmidt used could only have been invented by a man with both practical and theoretical experience. He had skilfully converted the disadvantage of the sagging of thin glass plates into an advantage for producing his corrector plate. Despite the enormous technical advances made in the last 50 years, his original process has continued to be of considerable importance in the production of corrector plates.

The Schmidt plate can be considered a lens and it is therefore

Fig. 3.4. Diagram a *shows (solid) the two zones that are ground from the glass plate as it is distorted by the vacuum (V) applied inside the chamber. A convex, spherical grinding tool is used. Diagram* b *shows the shape of the glass plate after air is again allowed to enter the chamber. This was the shape invented by Bernhard Schmidt to correct the spherical aberration caused by the spherical mirror. The illustration above clearly indicates this aberration.*

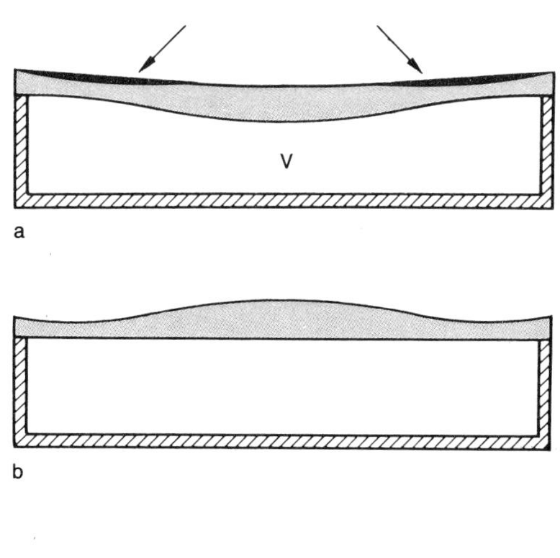

prone to the effects of chromatic aberration. Since the curvature of the different zones of the correcting plate is very slight, the chromatic aberration is very small and does not seriously affect the images. Bernhard Schmidt's ingenious invention showed that mirrors could be built having fields of view with diameters up to 15° across. The corrector plate is located at the centre of curvature of the primary mirror (Fig. 3.5). Dependent on the focal length of the primary mirror and on the diameter of the field to be covered, there is a relationship between the diameter of the primary mirror and the diameter of the corrector which serves as an aperture diaphragm. This relationship is determined by the fact that the whole of a bundle of rays having the maximum inclination to the optical axis should hit the primary mirror, and none of it be lost.

The system should function without any obstruction – i.e. it should be free from vignetting. The relationship between the mirror diameter (D_M), the corrector plate diameter (D_C), the focal length (F) of the mirror and the image field-diameter (α) is given mathematically by:

$$D_M = D_C + 4F \cdot \sin\frac{\alpha}{2}$$

The focal lengths amount to half the radius of curvature of the mirror. Several important conclusions can be drawn from this simple mathematical formula.

For a given mirror – i.e. with constant mirror diameter and focal length – and increasing image field, the correcting plate decreases, and with it the aperture ratio (the relationship between aperture and focal length continuously worsens). With an increase in the size of the field larger detectors are required in order to cover the whole field, and these, naturally, obstruct part of the primary mirror. These relationships can be explained with a specific example, which is shown numerically in Table 3.1.

The values given in the table are valid for a mirror with a diameter of 2 m and a focal length of 4 m. In the table, α is the diameter of the image field; F/D_C the reciprocal of the aperture ratio; B the length of the edge of the square image field, as given by the angle α; and F_B/F_C the ratio between the areas of the photographic plate and the corrector for normally incident light, showing the extent of the obstruction.

This example clearly shows that for a small field of view with a diameter of only 1°, a clear aperture of 186 cm is possible. The aim of the Schmidt telescope, however, is to obtain large-diameter fields of view. With a diameter of 10° the corrector plate can only measure 60 cm, and therefore the aperture of the telescope is greatly reduced. The size of the detector increases rapidly with increasing field diameter, rising from a diameter of 4.9 cm to 49.6 cm. With the largest vignetting-free field, the aperture ratio only amounts to 1 : 6.6, but what is decisive is that even for normal incidence the image area obstructs 68% of the clear aperture. In order to obtain the largest possible field of view, a diameter of 5° can be seen to be optimum. This gives an aperture stop with a diameter of 130 cm, the good aperture ratio of 1 : 3.1, and the image area of 24.7 cm by 24.7 cm obstructs only 3.6% of the aperture. The area of the skies covered by this Schmidt camera amounts to 12.5 square degrees. A reduction in the image field from 5° across the diagonal down to 3°, would result in a coverage of only 4.5 square degrees on the sky. The advantages would be that the aperture would be increased to 158 cm, and that the obstruction would be reduced to less than 1%. This example makes it very clear that the most convenient combination of the interdependent parameters has to be chosen for the telescope concerned.

Although the Schmidt system represents a very decisive advance in the development of the technology of astronomical telescopes, the disadvantages of the Schmidt camera must also be considered. One disadvantage is the curved image surface. As photographic

Table 3.1. *Parameters of Schmidt telescopes versus angular field of view*

α(°)	D_C (cm)	F/D_C	B (cm)	F_B/F_C
1	186	2.15	4.9	0.0007
3	158	2.52	14.8	0.0087
5	130	3.07	24.7	0.0361
7	102	3.90	34.6	0.1151
10	60	6.60	49.6	0.6834

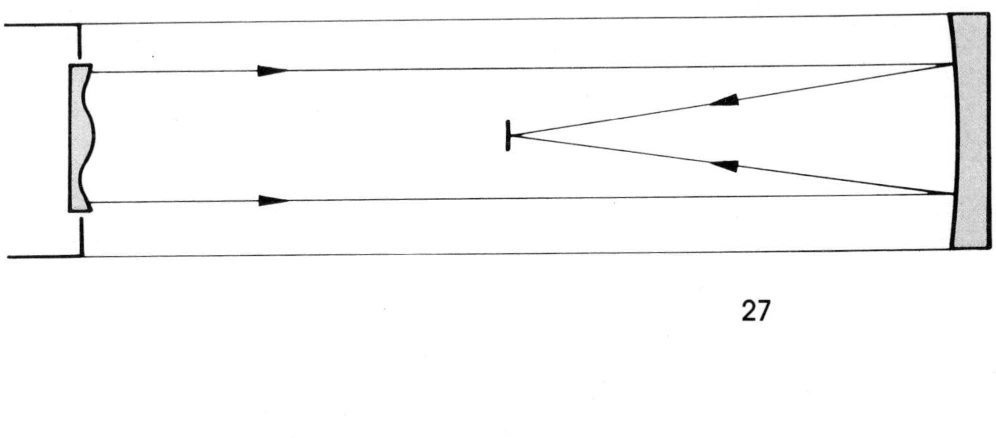

Fig. 3.5. The Schmidt system consists of two optical elements, a spherical mirror and a specially shaped corrector lens. The central area of the latter acts as a converging lens, and the outer zone as a diverging lens. It is located at the centre of curvature of the mirror. The foci of rays that enter the system at an angle to the optical axis fall on a curved focal surface. The radius of curvature of this focal surface is half that of the mirror and is equal to the focal length of the system.

plates are generally used in astronomical photography, the glass plates have to be bent to fit a spherical surface whose radius of curvature is the same as the focal length of the mirror. This makes great demands on the uniformity of the material from which the photographic plate is made. In practice, the focal lengths of the large Schmidt telescopes lie between 1.5 m and 4 m. In principle, it is, of course, possible to convert the curved image surface into a flat field by using a so-called 'field flattening' lens. However this immediately adds the problems of aberration that are associated with lenses. Therefore, the majority of Schmidt telescopes work with a curved image plane in order to make full use of the advantages characteristic of the coma-free Schmidt design. In some cases, as in Schmidt's original camera, for example, sheet film is used. This, however, gives rise to problems in fitting the film to the curved backing plate.

A Schmidt corrector plate behaves mechanically like the objective in a refractor. The cell can support it around its perimeter only, and not over the whole of the rear surface as with a mirror. Beyond a certain size, deformation and permanent distortion would result. This is the reason why no Schmidt telescope with an aperture larger than 150 cm has yet been manufactured, and why the instrument with a clear aperture of 134 cm that was installed at the then GDR's Academy of Sciences' Karl Schwarzschild Observatory in 1960 remains the largest (Fig. 3.6).

The crucial disadvantage of the Schmidt telescope, however, is the length required, which is at least double the focal length. The corrector plate, the primary mirror, and the image plane have to remain in a completely stable position relative to one another for every position of the telescope. This high stability, and the overall length required lead to a very massive telescope tube and this, in turn, demands an efficient mounting. The length of the telescope tube also requires a large dome. Specific figures for the largest Schmidt telescope in the world at present – corrector plate clear aperture of 134 cm – confirm this. The body of the tube alone has a mass of 28 tonnes. The total weight of the telescope tube and its stable fork mounting comes to 65 tonnes.

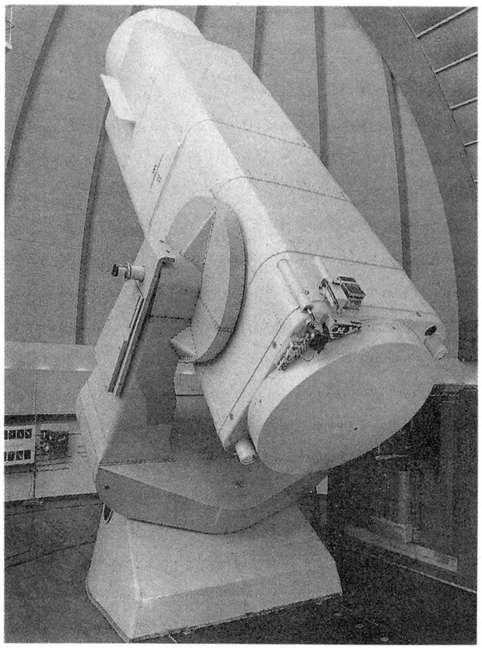

Fig. 3.6. The largest Schmidt telescope in the world has been in operation at the Karl Schwarzschild Observatory in Tautenburg since 19 October 1960. The telescope may also be used as a quasi-Cassegrain system, and a large spectrograph may be attached to one arm of the fork, where an eyepiece may be seen in this photograph. The guide telescopes for finding and subsequent guiding are located in two corners of the virtually square-section tube. A guide telescope eyepiece, a small finder and operating equipment can be seen at the rear of the tube.

One can only speculate on the dimensions of a Schmidt telescope with a corrector plate with a free aperture of 3 m. With a 3-m diameter corrector plate, and the fairly common aperture ratio of 1 : 3, the distance between the corrector plate and the mirror would be 18 m. With a field of view 5° in diameter, a primary-mirror diameter of 4.6 m would be necessary – or 4 m for just 3°. In order to cover a field 5° square, plates of at least 60 cm by 60 cm would be necessary, which would obstruct about 4% of the aperture. For a 3-degree field, photographic plates 45 cm by 45 cm would be needed. The distance of 18 m between the primary mirror and the corrector plate leads to an overall length of telescope tube of 21 m, including the mirror cell and the dew-cap. The tube of a Schmidt telescope with an aperture of 3 m is therefore almost as long as that of a 6-m telescope with a primary focal length of 24 m. Despite these enormous dimensions, however, the gain over the largest current telescope of this type would only be 1.75 magnitudes.

In spite of all its disadvantages, the Schmidt telescope remains a definite, significant achievement in the development of astronomical telescopes, and this is confirmed by the numerous, exceptionally important astrophysical results that have been obtained through its use.

Shortly after the ingenious idea of the coma-free Schmidt system became known, consideration was given to ways of overcoming its disadvantages, and, very soon, practical attempts were made towards this end. These attempts concentrated mainly on removing the great length inherent in the Schmidt system, without losing any of its advantages. Among the forms of reflecting telescopes, the Cassegrain type has the shortest overall size for a given focal length, which is what determines the size of refractors or a Schmidt system. This

Fig. 3.7. In the Schmidt–Cassegrain system the distance (CM) between the correction plate (C) and the primary mirror (M) can vary significantly. If CM is to be particularly small, i.e. if the smallest possible telescope is to be produced, then both mirrors deviate greatly from a spherical shape. The centres of curvature of both mirrors then lie in front of the correction plate and this also has to be strongly deformed. The length of the telescope can be reduced to just a third of the focal length. A classical Schmidt system with a focal length of 4 m has a tube 8 m long. With a Schmidt–Cassegrain system a focal length of 4 m can be obtained with an overall length of just 1.4 m. (F=Focus, S=Secondary mirror.)

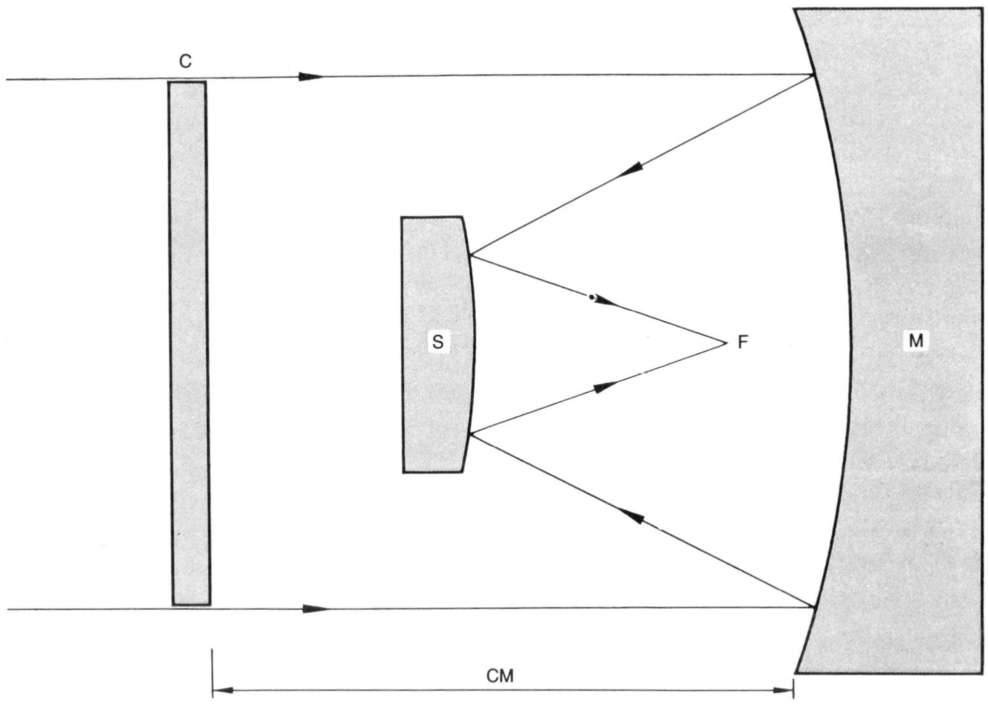

is possible because a convex, hyperbolic secondary mirror is located in front of the primary, just before the light reaches the focal point. This secondary increases the focal length. A Schmidt system with a focal length of 4 m has a tube length of 8 m and a refractor with a focal length of 8 m is also 8 m long. In contrast, a Cassegrain system (Fig. 3.7) with an equivalent focal length of 8 m can be built with a length of three metres. This example clearly shows the advantages of the Cassegrain system as far as the overall length is concerned. The focal plane in a Cassegrain system is situated behind the primary mirror which is perforated so that the light passes through it. With thoughts of reducing the overall length of the telescope, it is not surprising that attempts were made to marry the advantages of the Schmidt and Cassegrain systems. As early as 1940, James G. Baker used a convex secondary mirror in the Schmidt system in order to reduce the total, physical length as well as to flatten the image plane. A modified Schmidt system like this consists of a primary mirror to collect and focus the light, a corrector plate to produce a large field of view – it simultaneously serves as an aperture stop – and a secondary mirror to reduce the overall length and to flatten the field of view.

In constructing a Schmidt–Cassegrain system, the distance between the primary mirror and the corrector plate – which is rigidly determined in the Schmidt system – and the distance between the primary and secondary mirrors can be chosen freely. If a reduction in the length of the telescope is of prime importance, i.e. if the distance between the primary mirror and the corrector plate is to be particularly low, then the surfaces of both mirrors have to be aspheric. This is a crucial shortcoming because the production of aspheric surfaces has always been considerably more difficult than grinding and polishing spherical mirrors. In addition, the corrector plate has to be more strongly deformed in such a design than is required with the original Schmidt plate. This, of course, leads to a distinct chromatic aberration. One can attempt to correct this aberration by using a corrector plate made of different types of glass, or by using two corrector plates, as was attempted by E. G. Linfoot.

If the only aim is flattening the field then either the primary or the secondary mirror can be spherical, and the deviation of the other from a spherical form is also small. The telescope remains long, however.

A general disadvantage of the Schmidt–Cassegrain system is that the secondary mirror has to be relatively large, which means that the degree of obstruction is high. This example makes it very clear that it is not possible to retain the ingenious simplicity of the Schmidt system and, at the same time, eliminate its disadvantages.

A Schmidt–Cassegrain system was installed at the Boyden Observatory in Bloemfontein in 1950, and had the following parameters: diameter of primary mirror – 91 cm; diameter of corrector plate – 84 cm; diameter of secondary mirror – 40 cm; focal length – 220 cm; diameter of field of view – 7°. The large secondary mirror obstructs 23% of the clear aperture for normal-incidence light. With this system the smallest image diameters are 2 arc-seconds. This value exceeds the diameter of stellar images if the atmospheric conditions are good. With the original Schmidt system it is possible to produce image diameters of less than one arc-second without much difficulty.

In the Schmidt–Cassegrain system an additional optical component – the secondary mirror – was introduced in order to remove the Schmidt system's faults. Another method was described in 1944 by the Soviet optician, D. D. Maksutov (Fig. 3.8). He retained Bernhard Schmidt's basic principle of the spherical mirror, and simply chose another possible method of removing the spherical aberration. Maksutov corrected the spherical aberration by using a strongly curved lens, known as a meniscus. Similar systems were developed by others at approximately the same time. The crucial factor in Maksutov's system is that all the optical surfaces are spherical, which greatly simplifies their production. Maksutov applied for a patent for his idea as early as 1941.

The radii of curvature of the two surfaces of the meniscus generally differ only by a very insignificant extent, so the curved lens has almost the same thickness all over; causes minimal refraction strength and therefore introduces no chromatic aberration. The meniscus does, however, produce spherical aberration. This has to be equal and opposite to that of the spherical primary mirror, so that the two cancel each other out. Generally, the mirror and the meniscus have the same direction of curvature.

One very special system is the concentric meniscus system. Here, the centres of curvature of the meniscus surfaces and of the spherical mirror coincide. The centre of curvature of the curved image surface is also identical to those of the mirror and meniscus. All light-paths are equal, because there is no longer any specific optical axis, and large image fields may be attained. Unfortunately, chromatic aberration and residual spherical aberration cannot be avoided with the concentric meniscus telescope, although it is a very simple system.

When compared with the Schmidt plate, the meniscus has to be thicker, so, once again, it cannot be produced in large diameters. Maksutov was right in his assumption that a clear aperture of 100 cm is the absolute upper limit for meniscus telescopes. In fact, the largest meniscus telescopes have apertures of only 50 to 60 cm.

Inspired by Maksutov's publication the Schmidt system was even combined with the meniscus system. This combination, frequently used in astronomical observation, has been known since 1947. Two concentric meniscus lenses were combined with an achromatic corrector plate. This enabled an aperture ratio of 1 : 0.67 and an extremely wide field with a diameter of 52° to be attained. This optical system became known as the super-Schmidt camera (Fig. 3.9). The large angular field means that one can observe large areas of the sky, so it is particularly suitable for detecting objects whose position and time of appearance cannot be predicted. This means that super-Schmidt cameras are particularly employed for photographic observation of meteors. Although large aperture ratios and very large fields are possible with super-Schmidt cameras, large apertures are not. For this reason they can only be used for specific projects and not for general observations requiring a high light-grasp. The attempts to modify and develop the Schmidt system that have just been men-

Fig. 3.8. In this illustration of a Maksutov system, R_1 is the centre of curvature of the spherical primary mirror (M). R_2 and R_3I are the centres of curvature of the two surfaces of the meniscus. The meniscus lens (L) has hardly any effect on the rays entering in the centre but causes the outer rays to diverge. Their focal points then lie farther away from the mirror after they have been reflected than they would be without the meniscus lens. This produces a common focal point (F).

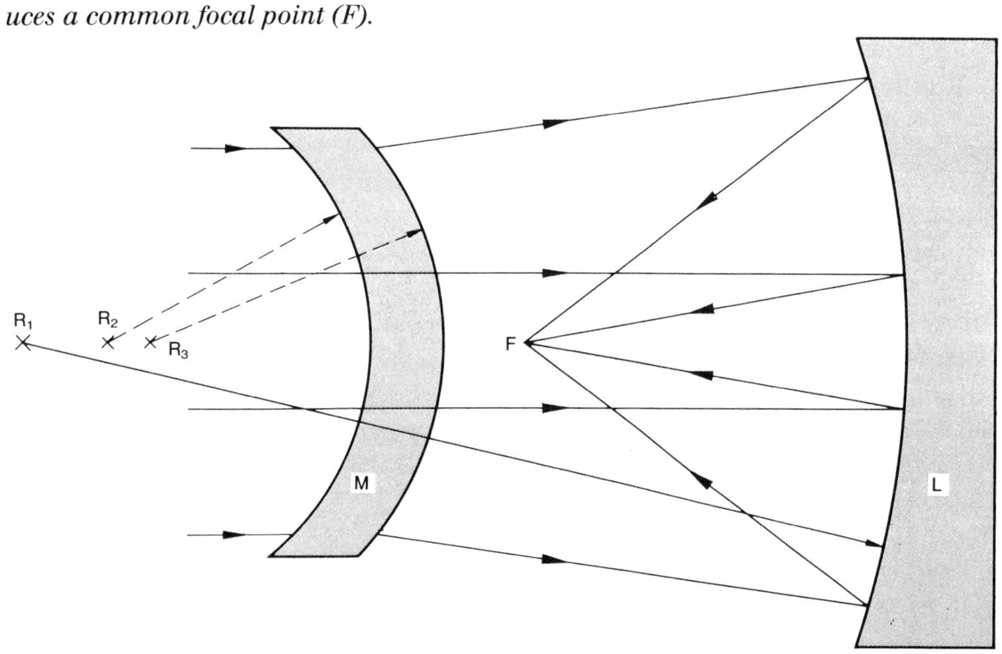

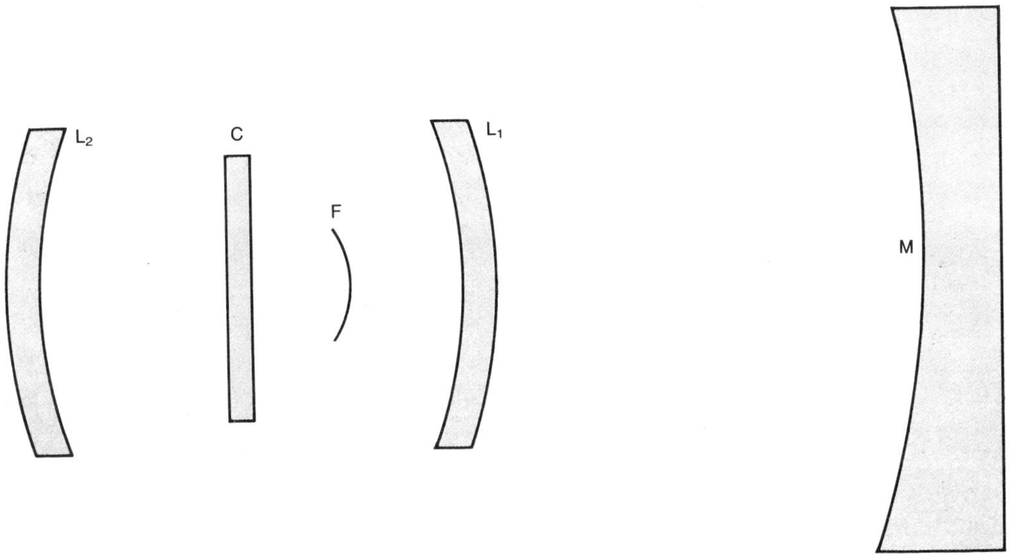

Fig. 3.9. With the original Schmidt system angular fields of view of up to 15° may be obtained. An extremely large field of view of 52° is obtained with the super-Schmidt camera. Two concentric meniscus lenses (L_1 and L_2) and an achromatic correction plate (C) are required in addition to the mirror (M). These optical components (C, L_1 and L_2) through which the rays of light have to pass, do, of course, lead to some loss of light.

tioned are only a selection, representing the various alternatives that were discovered. Apart from these, there is one very original variation, however, which is supposed to have been conceived by Schmidt himself. This has become known as the solid-Schmidt (Fig. 3.10). This consists of a glass block, whose front face is shaped like the corrector plate and whose rear face has to be spherically polished and coated. Here the light passes through glass right from the time it enters the Schmidt plate until it reaches the focus, reducing the focal length by the value of $1 : n$, and thus shortening the telescope. Here n is the refractive index of the glass, which lies between 1.5 and 1.7. It is immediately clear that the image surface is inaccessible, because it lies between the mirror and the corrector plate. D. O. Hendrix therefore suggested folding the system as shown in the diagram.

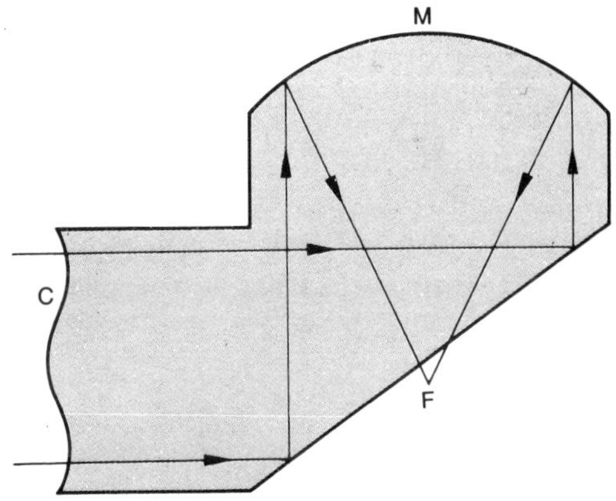

Fig. 3.10. The solid-Schmidt telescope was a very original idea. Such systems have been manufactured, giving focal ratios of up to f/0.3. In practice, solid-Schmidts had already been used in spectrographs by about 1940. It will be obvious that this variety cannot be produced in large apertures, because the mass increases very greatly as the size of the aperture increases. (C=Correcting surface, M=Mirror, F=Focus.)

Large Schmidt telescopes

The solid-Schmidt, like all the other modifications, has not found any general application in observational astronomy. Its uses have always been restricted to special applications. On the other hand, the original Schmidt system is used in many leading observatories.

Table 3.2 shows that 30 Schmidt telescopes with a corrector-plate diameter of more than 50 cm are in use. These details are taken from Landolt-Börnstein: *Zahlenwerte und Funktionen aus Naturwissenschaft und Technik*. The overwhelming majority of these telescopes, i.e. 23, were completed in the years 1951 to 1970. After 1970, only three more large Schmidt telescopes with apertures of over 100 cm were built. It is also interesting to note that one of the first Schmidt telescopes built after 1940 was also one of the largest ever constructed, with a clear aperture of more than 120 cm. Schmidt telescopes built and used by amateur astronomers have not been included in these statistics. It is quite conceivable that if they had been, the number of telescopes

Table 3.2. *Schmidt telescopes with a corrector-plate diameter of more than 50 cm completed in the years 1940–1986*

	Corrector plate diameter (cm)				
Year of completion	50–59	60–69	80–99	100–120	over 120
1940 to 1950		2	1		1
1951 to 1960	6	2	2		1
1961 to 1970	1	8		3	
1971 to 1980				2	1
1981 to 1986					

with apertures between 50 and 59 cm would be somewhat higher. Of the eight large Schmidt telescopes with corrector-plate apertures of more than 100 cm, seven were commissioned between the years 1960 and 1980 (Table 3.3).

Only three Schmidt telescopes have apertures which are significantly larger than 100 cm, and, of the eight largest Schmidt telescopes, only two are situated in the southern hemisphere. The use of large modern telescopes in the southern hemisphere is of particular interest. Because of the way in which institutions developed in the past, there were far more observatories in the northern hemisphere than in the southern, so significant gaps in observational coverage of the southern sky still remain. The number of Schmidt telescopes might be thought to be very low. It must be remembered however, that the time taken by observation with a Schmidt telescope is much less than the time necessary for evaluating the observational material. Schmidt telescopes are therefore very effective instruments. This point will be discussed in greater detail later in connection with observational programmes.

By the end of the 1940s, the first four Schmidt telescopes were ready for operation. In 1948, a 66/76 cm, *f*/2.9 Schmidt telescope began operation at Tonantzintla in Mexico; in 1950, a 61/91 cm, *f*/3.3 at Portage Lake for the University of Michigan – this instrument is now located at the Inter-American Observatory at Cerro Tololo in Chile – and in the same year, a 81/90 cm, *f*/2.6 Schmidt–Cassegrain system at Bloemfontein in South Africa. (The first two figures are the clear aperture and the diameter of the primary mirror respectively.)

One of the largest Schmidt telescopes ever constructed began operation at the very beginning of this period. This is the 122/183 cm (48/72-inch), *f*/2.5 system at Mount Palomar, which started operation in 1948. The mirror of Palomar's 'Big Schmidt' has a diameter of 183 cm, and edge thickness of 22.4 cm and a mass of 2.3 tonnes. The mirror rests on 36 support points in a welded steel cell. Eighteen of these are positioned according to the distribution of weight and act axially against the back of the mirror to counteract the flexure that would otherwise be caused by gravity. The other 18 are equally spaced around the perimeter of the mirror and therefore work in a radial direction. The distribution of the weight upon the two support systems depends on the position of the telescope. They compensate for the weight of the mirror in every position but they do not define the direction of the optical axis. So, a strong pin is situated in the centre of the mirror cell that fits into a perfectly centred hole in the back of the mirror. The mirror can therefore be radially adjusted in order to define the position of the optical axis.

In Schmidt systems, it is of great importance to maintain a constant value for the distance between the primary mirror and the image surface. This surface must always be located at a distance exactly equal to the focal length in front of the primary mirror. Any deviation from this entails a loss of image quality resulting in out-of-focus pictures. In the 'Big Schmidt' the distance between the plate holder and the mirror is determined by three Invar nickel-alloy rods with a diameter of 2.5 cm. The Invar material has a very low coefficient of thermal expansion and does not alter in length with temperature changes over the range of temperatures normally encountered. One end of each Invar rod is fastened to the

Table 3.3. *Schmidt telescopes with a corrector-plate diameter of more than 100 cm*

	Year commissioned	*d*/*D*/*F* (cm)	Latitude of its site in degrees
Karl Schwarzschild Observatory, Tautenburg, Germany	1960	134/200/400	+51
Mount Palomar Observatory, USA	1948	125/183/307	+33
Siding Spring Observatory, Australia	1973	124/183/307	−31
Kiso Observatory, Japan	1974	105/150/325	+36
Bjurakan Observatory, USSR	1961	100/150/213	+40
Kvistaberg Station, Uppsala, Sweden	1964	100/135/300	+60
European Southern Observatory, Chile	1969	100/160/306	−29
Llano del Hato Observatory, Venezuela	1978	100/152/300	+9

d: diameter of the corrector plate
D: diameter of the mirror
F: focal length

photographic plate holder at the focal surface in the centre of the tube. The other ends rest directly on the mirror. Because of the supports behind the mirror, the mirror always presses against the three Invar rods and consequently, against the plate holder. This ensures that the distance between the image plane and the mirror remains constant.

These methods of mounting the mirror and of fixing the distance between the mirror and the image plane that were chosen for the 'Big Schmidt' were employed in practically all subsequent large Schmidt telescopes.

The corrector plate of the Palomar Schmidt telescope has a diameter of 134.6 cm (60″). This leaves an effective clear aperture of 125.7 cm (50″) because the lens has to be supported around the perimeter. The corrector plate is only 9.5 mm thick. The neutral zone of the plate, i.e. the transition area between the central converging lens, and the outer diverging lens is situated 50.8 cm from the centre.

Remarkably different sizes for this corrector plate are to be found in the literature. Most of them seem to be based on the commonly used, but incorrect, designation of this telescope as the '48-inch Schmidt'. The values quoted here are from an authentic description by R. G. Harrington, where the dimensions are given in both inches and in metric equivalents.

As the corrector plate is a lens, chromatic aberration is produced. This is extremely slight over the whole range of optical wavelengths but it can only equal zero for one particular wavelength. The corrector plate of the 'Big Schmidt' is corrected for the wavelength of 440 nm, i.e. the image diameters are smallest for light at this wavelength. Because the focal length for light of other frequencies is slightly different, the image-diameters for light at other frequencies are slightly larger.

Depending on its composition, glass has certain transmission properties, which vary with the wavelength of the light. In the Big Schmidt's corrector plate the transmission of light decreases below 360 nm and at 340 nm reaches no more than 40%. Below 320 nm, the Schmidt plate is no longer transparent. With a focal length of 307 cm, 1 mm at the focal plane of the 'Big Schmidt' corresponds to 67.19 seconds of arc on the sky or, to put it another way, 1 second of arc is 14.9 μm. Two sizes of photographic plate can be used in the Palomar Schmidt telescope. With 36 cm × 36 cm plates the field of view amounts to 6.6° × 6.6°; with 24 cm × 24 cm the field is only 4.7° × 4.7°. Because of the large image field, some rays of light that are inclined to the optical axis suffer obstruction, causing what is known as vignetting.

The 'Big Schmidt' works like most large Schmidt telescopes with a curved image field. This means that the photographic plates have to be curved over the plate-holder during the exposure. The slightest of cracks in the glass, the smallest splintering at the edge caused when it was originally cut or any inhomogeneities in the material could lead to breakages in the plate-holder. As this is generally realized only after the exposure, the time taken by the observation is normally lost. In order to prevent this happening, at Palomar the photographic plates are bent to a radius of 254 cm in a special plate holder for one minute before exposure. This radius of curvature is smaller than the one in the telescope which is 307 cm. Only plates that pass this test are used for actual observation.

The photographic plate holder is automatically loaded into the focal plane from outside the tube. The cover that protects the plate from the light is then retracted, but, at this stage, the mirror and the photographic plate are not yet exposed to the light. The shutter controlling the exposure is situated immediately behind the corrector plate. This shutter has an aperture of 125.7 cm, and it takes 3 seconds to open or close. It is completely vibration-free. The 'Big Schmidt' is fitted with two guide telescopes with apertures of 25.4 cm and focal lengths of 381 cm. The whole telescope is carried and moved by an equatorial fork mounting. The movement of the telescope, the dome and the slit, etc, can be operated and controlled from a central control desk.

Twenty-five years after the 'Big Schmidt' was completed, another telescope with the same optical parameters was built. It is operated by the Royal Observatory, Edinburgh, and is sited at the Siding Spring Observatory near Coonabarabran in New South Wales in Australia.

The optical parameters of the UK (United Kingdom) Schmidt are practically identical to those of the 'Big Schmidt' – mirror diameter 183 cm; aperture of the corrector plate 124 cm; focal length 307 cm; size of photographic plate 35.6 cm × 35.6 cm; field of view 6.4° × 6.4°; plate scale one arc-second equals 14.9 μm. As mentioned for the 'Big Schmidt', the large plate-size is affected by vignetting. The experimentally determined vignetting function of the UK Schmidt shows that the light-loss begins a short distance away from the centre of the plate, in accordance with the theory of geometrical optics (Fig. 3.11). One can see from the diagram that the apparent brightness near the edge of the plate is reduced by approximately 0.2 magnitudes as a result of the vignetting.

Since 1977 the UK Schmidt has had a new correction plate made from cemented UBK7 and LLF6 glass. The choice of two types of glass for the Schmidt plate removes nearly all chromatic aberration. It is therefore possible to attain an image diameter over the whole range of optical wavelengths of less than one arc-second. The first corrector plate was corrected for 440 nm but gave large diameters in the long-wave region. The transmission of the corrector plate for a range of wavelengths is shown in Table 3.4.

The UK Schmidt telescope is equipped with two objective prisms for the production of spectra. In order that as many spectral details as possible are recognizable, the reciprocal dispersion should have a high value, and this is obtained by using a prism with a large angle of refraction. With long spectra, there is the danger that spectra of neighbouring objects will overlap. In addition, one of the aims is to examine faint astronomical objects spectroscopically. This requires a lower reciprocal dispersion. In order to meet these incompatible tasks, two prisms were made for the UK Schmidt. The first, prism A, has a refractive angle of 44 arc-minutes and the other, prism B, one of 2.25°. This means that quite different dispersions can be produced, which, for each prism, depends on the wavelength. With the first prism the lengths of the spectra from 300 nm to 700 nm amount to about 1 mm, and with the second, nearly 3 mm. The prisms can be used individually or together when they can be mounted so that their dispersions either combine (AB) or are opposed to one another (BA). The reciprocal dispersions resulting from their combined use are also given in Table 3.5.

Fig. 3.11. As the angle between the incident pencil of light and the optical axis increases, so the pencil of light becomes more and more vignetted. Because of this vignetting, objects at the edge of the field appear fainter. In astronomical terms this means that the brightness of an object decreases as its angular distance from the optical axis increases. In the UK Schmidt vignetting has a noticeable effect on objects at an angular distance of more than 2° from the optical axis. At 4° the vignetting leads to a loss of about 0.2 magnitudes. If its dependance on angular distance from the optical axis is known, the observations can later be corrected.

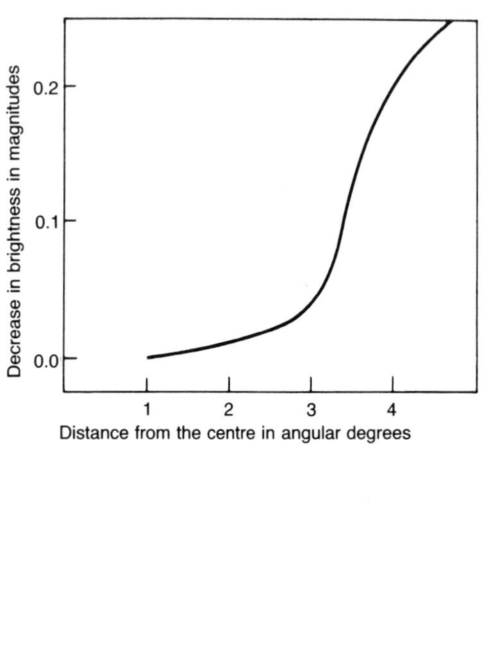

Table 3.4. *The relationship between transmission and wavelength of the UK Schmidt*

Wavelength in nm	310	320	330	340	350	370	390	420
Transmission (%)	9	39	56	74	82	90	94	98

The UK Schmidt also has a small prism with an aperture of only 25 cm. If this prism is placed in front of the corrector plate, it produces a second weaker image alongside the main image. Because of the ratio between the diameter of the corrector plate and the aperture of the prism, the reduction amounts to 3.1 magnitudes. If parts of the prism are blocked off, leaving an aperture of only 5 cm, then the reduction reaches 6.6 magnitudes. The secondary images lie 24 arc-seconds, or 0.362 mm, away from the main image. If one knows the apparent magnitude of a particular star, the secondary images will have well-defined fainter magnitudes. The calibration curve – which is required to determine the magnitudes of previously unstudied stars – can be extended 3.1 or 6.6 magnitudes below the faintest magnitude otherwise available. This is very important for carrying out photometric work with Schmidt photographs, because accurate, photoelectrically determined, magnitudes do not exist at the faint magnitudes that can be reached photographically with Schmidt telescopes.

In the international spectral bands B (blue), V (visual) and R (red), the UK Schmidt's limiting magnitude is quoted as being 21. Under good atmospheric conditions, the faintest magnitudes are reached after an exposure time of 60 minutes. The limits are not, of course, so faint for photographs with objective prisms. When using the Kodak IIIa–J photographic plate, a magnitude of 20 is reached with the 44 arc-second-prism after an exposure of 45 minutes, and a magnitude of 18 with the 2.25°-prism.

The UK Schmidt is one of the few Schmidt telescopes in the southern hemisphere. The largest Schmidt telescope in the world has been in operation since 1960, at the Karl

Table 3.5. *The reciprocal dispersions versus wavelength for both prisms of the UK Schmidt telescope*

Wavelength in nm	Reciprocal linear dispersion in nm per mm A	B	AB	BA
656.3 (Hα)	809	275	205	417
486.1 (Hβ)	347	118	88	179
434.0 (Gγ)	244	83	62	126
397.0 (Hε)	185	63	47	96

Schwarzschild Observatory (KSO) (Fig. 3.13). The aperture of the corrector plate is 134 cm; the diameter of the mirror 200 cm; the focal length 400 cm, and the field of view 3.4° × 3.4°. This Schmidt camera works with photographic plates 24 cm × 24 cm, and during the observation these are carved to fit a spherical surface with a radius of 400 cm. On the photographic plate, 19.4 μm correspond to one arc-second. There is no vignetting.

We have already mentioned that the corrector plate of the 'Big Schmidt' is correct for wavelengths of 440 nm. The corrector plate of the KSO Schmidt, by comparison reaches its smallest image sizes at 630 nm. This is clearly shown by the percentage of light that is concentrated in an area of a particular diameter (see Table 3.6). It can be clearly seen how the image size improves (i.e. decreases) towards longer wavelengths. Figure 3.12 shows image diameters for the three largest Schmidt telescopes. In the long-wave region the advantage of the KSO Schmidt over the 'Big Schmidt' is obvious. In the 440 nm region the image size is better in the 'Big Schmidt'. The illustration also clearly illustrates the advantage of a corrector plate made from two types of glass, i.e. an achromatic plate as was made later for the UK Schmidt. The original corrector plate, made of BK7 glass gave large image-diameters particularly in the long-wave region.

The KSO Schmidt (Fig. 3.13), like many other Schmidt telescopes, is equipped with an objective prism. The use of objective prisms, especially large ones with diameters of 100 cm and over, causes problems. The large prisms affect the stability of the telescope because of their extra weight. In addition, reflections arise at the surfaces of a prism, just as they do at the surfaces of the corrector plate, and these produce unwanted secondary images of bright stars, known as ghost images. Astronomers, especially those who work with the UK Schmidt telescope, have investigated these problems, because two prisms in front of the corrector plate naturally increase the number of possible ghost images (Fig. 3.14). Reflections occur at both the front and rear surfaces of the prisms. When using one prism, three ghost images appear caused by the corrector plate and the prism: a reflection from the rear of the prism onto its front surface and thence to the mirror, and two reflections from the corrector plate, one onto the rear surface of the prism, and the other onto the front, both then being returned towards the mirror. When using two prisms the number of ghost images increases to as many as ten. This does not take into consideration the fact that light can also be reflected from the emulsion or a filter onto the corrector plate and then onto the mirror. Fortunately, multiple ghost images from a star occur very rarely because only with bright objects is the intensity of the reflected light strong enough to give rise to a clearly visible effect on the photographic plate.

The photograph of the well-known Pleiades cluster taken by the Tautenburg Schmidt (Fig. 3.14)

Table 3.6. *Spot concentration (%) in the KSO Schmidt telescope for different wavelength ranges*

Diameter in arc-secs	360 nm to 400 nm	490 nm to 650 nm	580 nm to 670 nm
1.2	67	88	100

Fig. 3.12. The image diameters of the three largest Schmidt telescopes show a different dependence on wavelength. The largest Schmidt telescope in the world (KSO) has its smallest image diameter in the red spectral region. In the 'Big Schmidt' at the Palomar Observatory, radiation at 430 nm produces the smallest stellar images. The optimum value of the original corrector plate of the UK Schmidt telescope (UK1) lay in the same region. The second corrector plate used in this instrument (UK2) is made of two types of glass and produces small-diameter images that are almost independent of wavelength. The angular diameters given refer only to the optics of the three telescopes themselves.

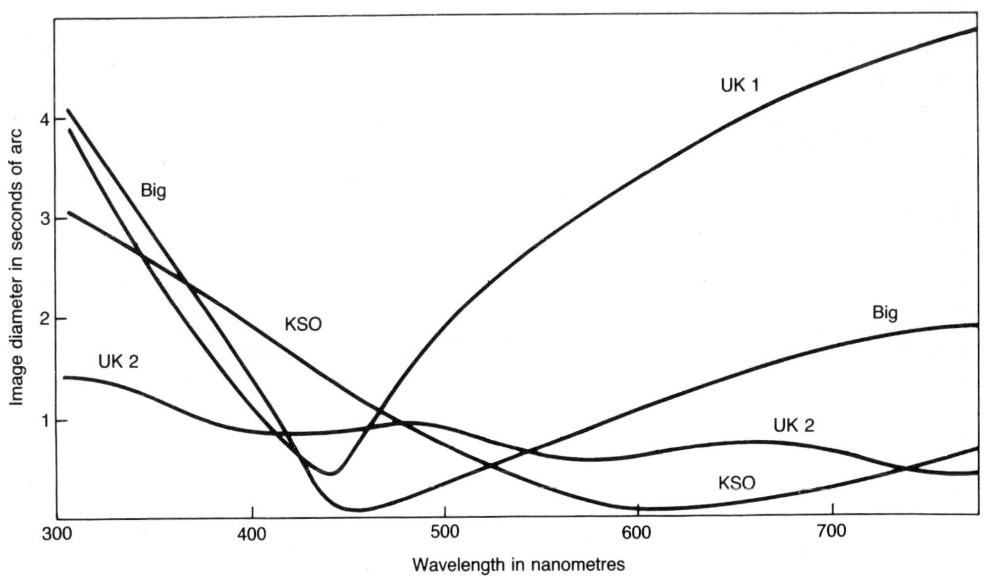

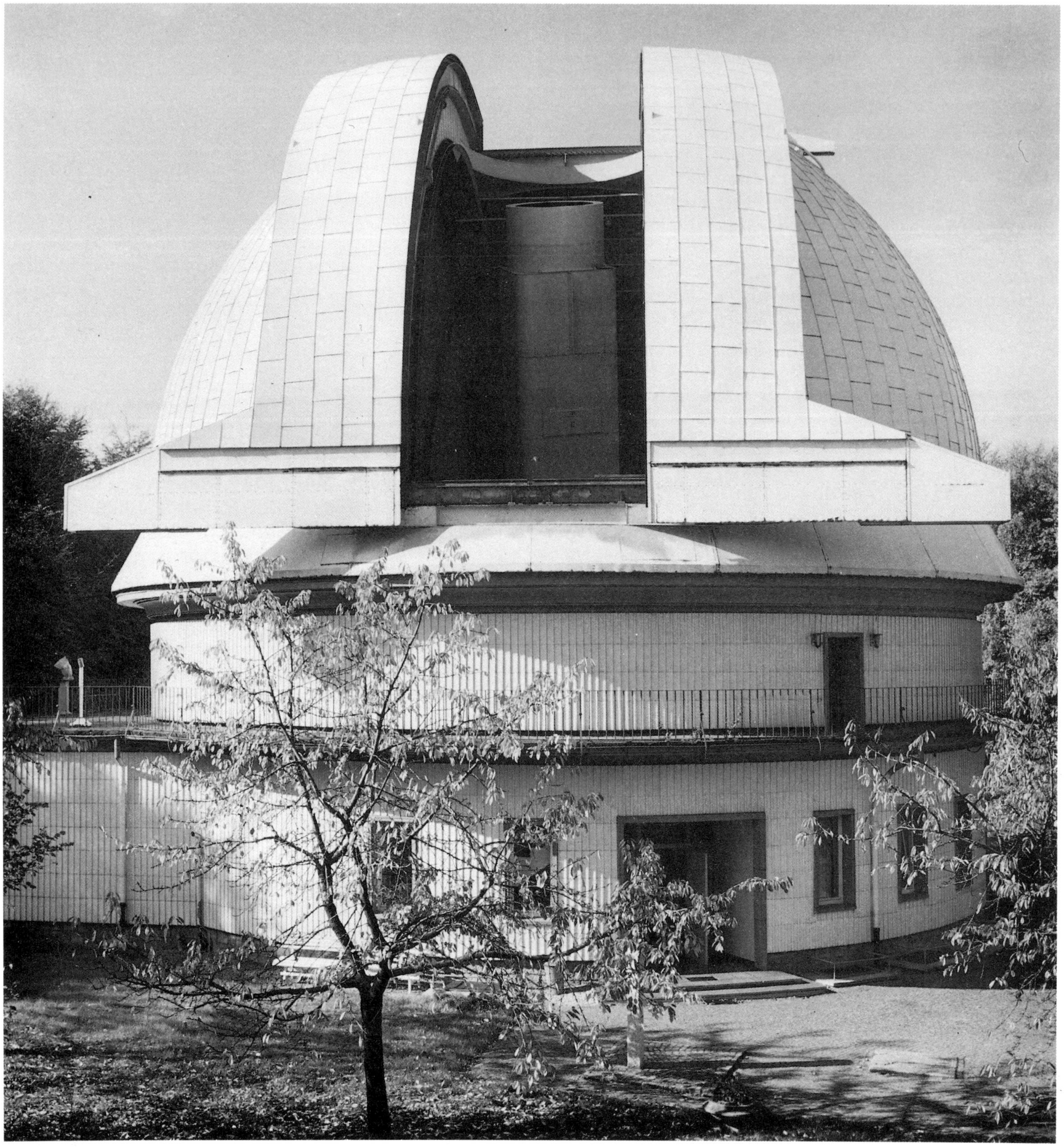

Fig. 3.13. The Schmidt telescope dome at the Karl Schwarzschild Observatory at Tautenburg. A typical form of protection for telescopes is the rotating dome. For Schmidt telescopes with a clear aperture of 1 m and above, domes with diameters of up to 20 m are necessary. The size of the dome depends on the length of the telescope, which in turn depends on the focal length of the system. The KSO Schmidt telescope has an aperture of 134 cm, a focal ratio of f/3 and thus a focal length of 4 m. The length of the telescope tube is 8 m, not including the cover and the dew-cap,

clearly shows ghost images of the brighter stars. On the original plate these ghosts are not so distinct because their photographic density is not far above that of the background. But this is where the danger lies, because the darkening caused by the ghost images naturally affects photometric measurements of stars covered by the ghost images. The 'Ghost Pleiades' can be clearly recognized in the picture because a special procedure was used to enhance faint contrasts. The ghosts are extremely out-of-focus images of stars and have correspondingly large diameters. Within the ghost images the central plate-holder, its four supports and the surrounding telescope tube can all be detected.

In order to save weight, but more particularly to avoid numerous ghost images, as early as 1956 the suggestion appeared in the literature that the corrector plate and the objective prism should be made in one piece. Carl Zeiss Jena were the first to put this suggestion into practice, and so far have remained the only firm to do so. As no one had any previous experience of producing a prismatic corrector plate, before a large plate was cut for the KSO Schmidt, a smaller one with a diameter of 60 cm was made for the 60/90 cm, *f*/3 Schmidt telescope at the Jena University Observatory. The prismatic side of this Schmidt plate has a refractive angle of 1°, which together with the 180 cm focal length of the Schmidt camera, gives a reciprocal dispersion of 250 nm per millimetre at the wavelength of the Hγ-line. The test results of this prototype plate were exceptional, and so Carl Zeiss produced a prismatic corrector plate with a diameter of 140 cm and a refractive angle of 0.5° for the KSO Schmidt. With a focal length of 4 m, this also gives a reciprocal linear dispersion of 250 nm per millimetre for Hγ. A detailed examination of limiting magnitudes showed that the magnitudes of stars that could be identified were as follows: Kodak 103a-O – magnitude 20.4; Kodak

Caption for fig. 3.13 (*cont.*)
the round cylinder visible through the slit in the dome. The approximate diameter of a dome may be taken as being slightly more than four times the focal length of the Schmidt system.

Fig. 3.14. In this photo of the Pleiades the ghost images can be attributed to the bright stars as indicated by the lines. The lines connecting the direct images with the corresponding ghost images have a single point of intersection.

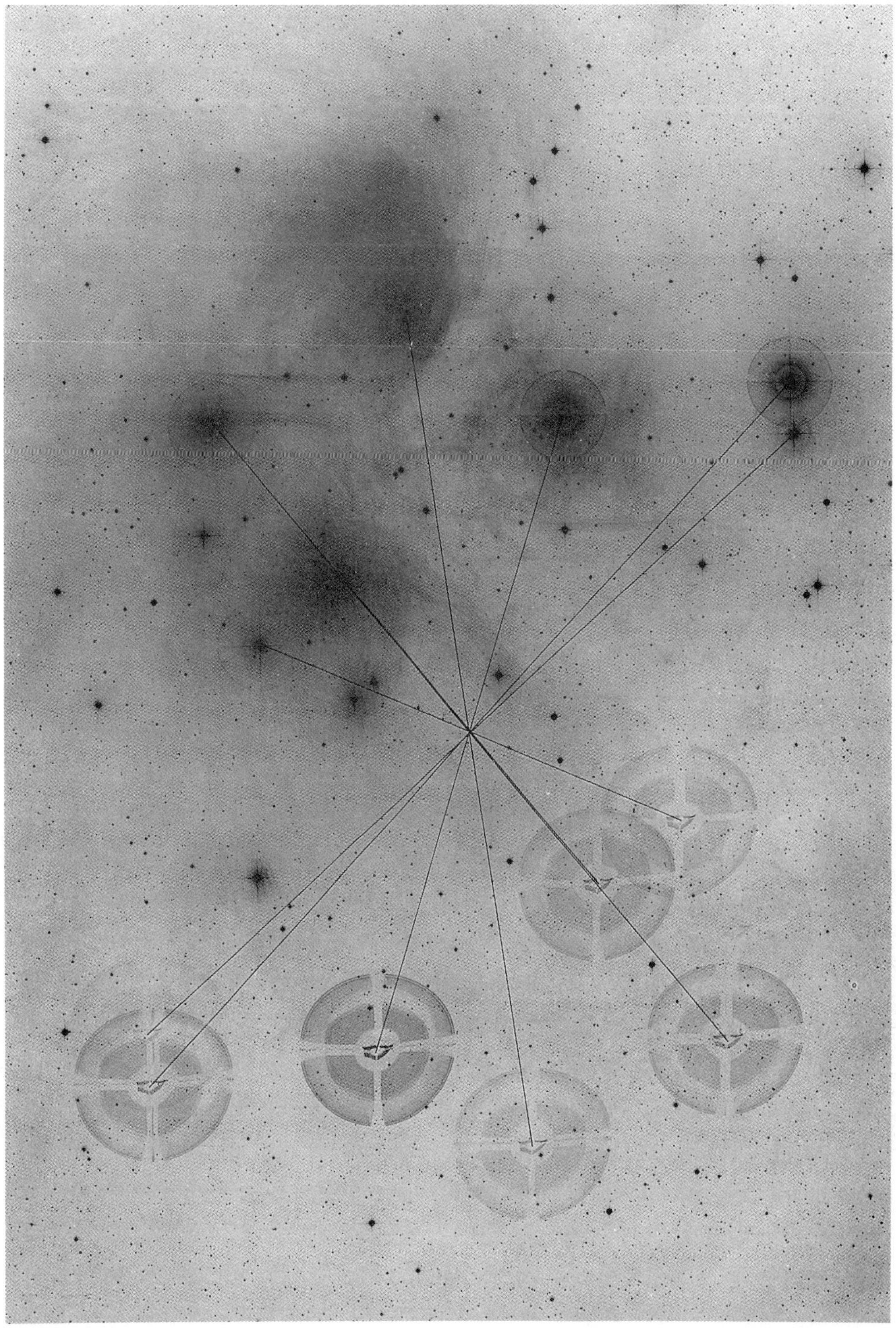

103a-D – magnitude 19.0; Orwo NO 27 – magnitude 18.5.

The Karl Schwarzschild Observatory's two-metre telescope is often described as a universal reflector because it can also be used as a Cassegrain (21 m focal length), and as a Coudé system (92 m focal length), although naturally with only a small field of view. This telescope is therefore very adaptable and can also be used for photoelectric and spectroscopic observation of individual objects. In practice, the Tautenburg 2 m telescope is regularly used for two weeks as a Schmidt camera, and then for two weeks as a Cassegrain, or a Coudé system. The reason for this is that instruments with large aperture ratios cannot be used effectively for observations when there is a very bright sky background. The brightness of the background largely determines the signal-to-noise ratio. This means that the background limits the maximum useful exposure time. Greatly simplified, it can be imagined as follows: the photons coming from the sky background cause a darkening of the whole photographic plate. The optimum signal-to-noise ratio is reached with a very specific background level. If the sky background is very dark, i.e. if there are few background photons, one can expose the photographic plate for a long time. In other words, the time over which photons from celestial objects (the signal photons) can 'build up' is long, and very faint objects can still be detected. If the sky background is very bright, then the photographic plate's optimum background level is reached after only a short time, within which only bright objects are recorded. The best exposure time, and thus the optimum limit of Schmidt telescopes is therefore determined by the brightness of the background sky. The natural night sky is very bright a few days before and after full Moon and therefore observations with Schmidt telescopes are not worthwhile during this period, because only short exposures would be possible. So the capacity to use the Two-Metre telescope at the Karl Schwarzschild Observatory not just as a Schmidt camera, but also as a Cassegrain or as a Coudé system with a very small *f*-ratio – particularly for spectroscopic work, which is not affected by a bright sky background – represents an ideal solution. In order to be able to determine the best exposure time for the KSO Schmidt objectively, the sky background is continuously monitored by a photoelectric photometer (Fig. 3.15). This photometer works over the whole range of wavelengths at which photographic observations can be made with the Schmidt camera. Apart from the

Table 3.7. *Schmidt telescopes with corrector-plate diameters of more than 60 cm*

Observatory	Diameter (cm) CP	M	Focal length (cm)	Year commissioned
Ann Arbor	61	91	214	1950
Asiago	65	92	215	1964
Budapest	60	90	180	1963
Bjurakan	100	150	213	1961
Bloemfontein	81	90	303	1950(1)
Calar Alto	80	120	240	1980(2)
Castel Gondolfo (Vatican Obs.)	64	93	240	1976
Cerro La Silla (ESO)	100	160	306	1969
Cerro Tololo (Interamerican Obs.)	61	91	213	1967
Cleveland	61	91	214	1957
Coonabarabran (Siding Spring Obs.)	120	180	306	1973
Gran Sasso (Rome Obs.)	65	95	190	1959
Hsing-lung (Peking Obs.)	60	90	180	1963
Jena	60	90	180	1963
Kiso Obs.	105	150	325	1974
Kvistaberg (Uppsala Obs.)	100	135	300	1964
Llano del Hato (Uni. of Andes, Merida)	100	152	300	1978(3)
Mount Palomar	122	183	307	1948
Saltsjöbaden (Uni. Stockholm)	65	100	300	1964
Tautenburg (Karl Schwarzschild Obs.)	134	200	400	1960
Tonantzintla	66	76	217	1948
Toruń	60	90	180	1962
Uccle	84	120	210	1958

CP = Correction Plate M = Mirror
(1) Baker Schmidt–Cassegrain system
(2) Previous site Hamburg–Bergedorf
(3) Previous site Caracas
Source: This table was compiled from: *Landolt–Börnstein: Zahlenwerte und Funktionen aus Naturwissenschaft und Technik* (Springer-Verlag), Neue Serie (Gruppe IV) 1965 and Neue Serie (Gruppe IV Band 2a) 1981

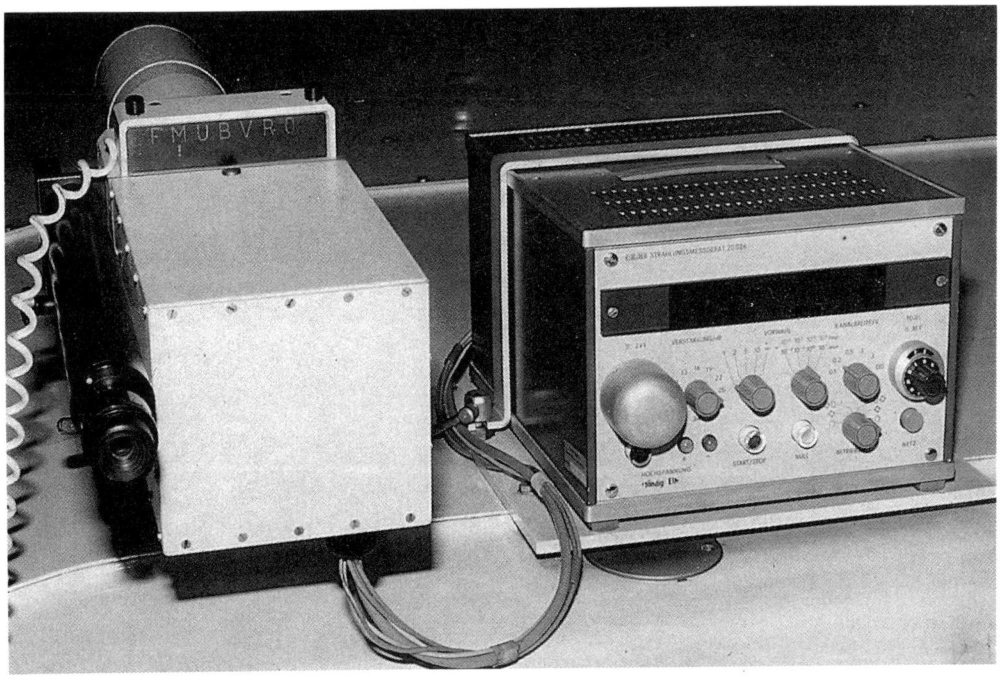

Fig. 3.15. The exposure meter used with the KSO Schmidt telescope at Tautenburg is a photoelectric photometer, which measures the number of photons coming from the sky background. The photometer can be moved slightly to avoid any readings being distorted by the presence of a bright star in the field of view. Standard curves exist for the various colour bands, showing the relationship between a specific numerical value and the resulting darkening of the background on a photographic plate.

three mentioned here, the most important Schmidt telescopes are included in Table 3.7. Their optical and mechanical construction and their equipment are essentially the same as those described here.

4 Astronomical applications of Schmidt telescopes

The observational tasks assigned to Schmidt telescopes result from this optical system's large field of view. A basic rule for classical, parabolic-mirror systems is that the diameter of the field of view in minutes of arc corresponds to the aperture ratio, i.e. a classical Cassegrain telescope with an aperture of 1.4 m and a focal length of 14 m has a field diameter of 10′. By using additional, optical corrector systems, usable fields with a maximum diameter of 1° can be achieved. With Schmidt reflectors, by comparison, it is easily possible to obtain coma-free fields from 3° to over 5° in diameter.

From this characteristic property of the Schmidt telescope specific research tasks can be derived. In this respect conclusions were reached at the International Astronomical Union Colloquium 'Astronomy with Schmidt telescopes' that was held at Asiago Observatory in Italy in 1983, and at which various categories of objects that are particularly suitable for observation with Schmidt telescopes were compiled.

The field of large, ordinary telescopes amounts to one square degree at most. Numerous astronomers want to observe with large telescopes and, as a result, only limited time is available for any particular observing programme. In general, it is not possible to record more than ten square degrees of sky in the course of one programme. If we assume that, in order to obtain sufficient information about a particular group of objects, at least 25 objects have to be recorded, on average, therefore, there would have to be 2.5 objects per square degree. That means that, over the whole sky, there would have to be 100,000 members of the group under investigation. If there are less than 100,000 objects in a particular class, then telescopes with large fields, i.e. Schmidt telescopes, have to be used.

In the light of this, the following classes of object are especially suited for observation with Schmidt telescopes:

- galaxies brighter than 16th magnitude;
- peculiar galaxies;
- quasars brighter than 19th magnitude;
- clusters of galaxies to a distance of 1500 Mpc;
- superclusters to a distance of 2400 Mpc;
- peculiar stars brighter than 15th magnitude;
- variable stars;
- minor planets;
- comets.

Schmidt telescopes, with their large fields, also have to be used for the observation of extended objects, particularly including interstellar matter, which often appears in the form of large, luminous, gaseous nebulae or of dark clouds, or in the special form of expanding supernova remnants. Such objects are often well over 1° in extent.

In recent years, and still today, astronomers have been concerned to map the whole sky by photographic means. If such exposures are to be obtained at two different wavelength regions, using a large telescope with a field of one square degree, it would require 85 000 exposures. With a Schmidt telescope, with a field of 12 square degrees, 7000 exposures are enough. If 1 h is required for each exposure, and at one observation site 1200 hours are available per year, then the task can be completed in three years with one Schmidt telescope in the northern, and one in the southern hemisphere. Even if we assume that, because of its larger aperture when compared with a Schmidt telescope, a large conventional telescope could reach the same limiting magnitude with an exposure of only 30 minutes, with two telescopes each working 1200 hours per year, 18 years would be required to map the whole sky photographically. This simple example also clearly illustrates the type of task for which Schmidt telescopes are particularly suitable.

Mapping the sky

The most important scientific programmes for Schmidt telescopes have arisen from considerations outlined in the previous chapter; some of them may be briefly discussed. Several years after the completion of the 'Big Schmidt' at the Palomar Observatory, it was used for the first extensive programme of mapping the northern sky. It was planned to photograph the entire northern sky down to declination −27°, in both the blue and red spectral regions. Experience gathered during this programme led to the photographic survey of the northern sky being extended down to declination −33°, although a slight reduction in the quality of this additional zone could not, of course, be avoided. A total of 935 fields in two colours were taken for the *Palomar Observatory Sky Survey.* The centres of the fields were chosen so that neighbouring fields overlapped by 0.6°. The exposures for the survey were taken only on nights with good seeing and without moonlight. The diameter of the seeing disc was not to be greater than 3 seconds of arc. Only for the zone at −30° declination could this condition not be met. All exposures were produced within two hours of the particular field's meridian transit, in order to keep the effects of extinction, refraction and instrumental flexure as low as possible.

For the exposures in the blue spectral region, Kodak 103a-O emulsion was chosen. The red spectral region can be achieved with Kodak 103a-E emulsion and a Kodak 2444 plexiglass filter. The blue plates required exposure times of 10 to 15 minutes, and for the red, 40 to 60 minutes. This led to limiting magnitudes of 21 in the blue, and 20 in the red (Fig. 4.1).

This first photographic survey of the sky to be made with a Schmidt telescope has been the basis of important scientific investigations at many astronomical institutions. A total of 200 copies of the original plates were produced and sold to other institutes, and there is hardly any field of astronomy that has not benefited from the *Palomar Observatory Sky Survey.*

The thought of extending this important photograph atlas to cover the whole southern sky was therefore proposed and this was carried out by the ESO Schmidt and the UK Schmidt, both of which are in the southern hemisphere. In 1973 and 1974 606 fields analogous to the 'Palomar Survey' were photographed with the ESO Schmidt on Kodak IIa-O-plates, covering the declination zone from −17° to −90°. This series of exposures is known by the classification ESO(B), since it covered the blue spectral region. For the ESO(R) plates in the red wavelength region, which were exposed subsequently, the new, more sensitive, Kodak 098-04 emulsion was chosen in preference to Kodak 103a-E.

With a collaborative programme, the ESO(B) and ESO(R) plates were supplemented by a set of exposures on Kodak IIIa-J plates carried out with the UK Schmidt in Australia. These 606 maps of the sky are known as SERC(J) (Science and Engineering Research Council). By 1980, this extensive programme was eventually essentially complete. Some 170 astronomical institutes now use copies of the *ESO/SERC Survey.*

Fig. 4.1. The Palomar Observatory Sky Survey was carried out in two wavelength regions. The diagram emphasizes that the width in the blue spectral region (175 nm) is much greater than in the red (100 nm). This was also a reason for using longer exposures in the red spectral region. On the blue plates all the objects with a radiation curve peaking in the short-wave spectral region are prominent. These are, for example, early-type hot stars. On the red exposures late-type cool stars and stars reddened by interstellar matter are easier to recognize.

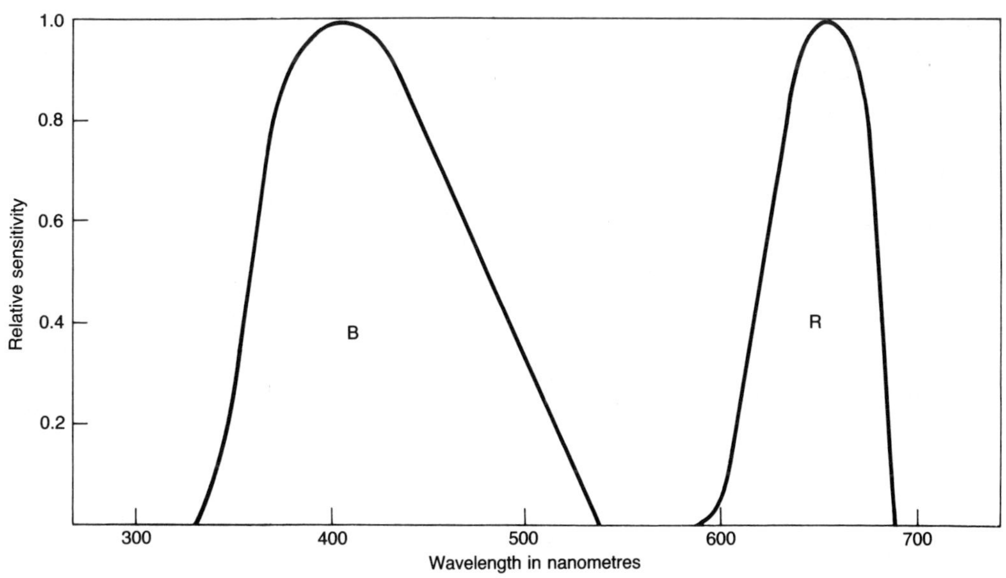

The UK Schmidt then began an infrared survey of the southern sky, designed to support the work carried out with the UK 3.8-m infrared telescope (UKIRT) in particular, and also to extend the data collected by the IRAS satellite. The infrared exposures on Kodak IV-N plates with an RG 175 filter can also be obtained when the Moon is bright, and therefore do not interfere with any other programmes. In the first phase, 163 fields were taken in a band along the southern galactic equator and of the area around the Magellanic Clouds. The infrared plates, which required an exposure of 90 minutes, were supplemented by a set of Kodak IIIa-F plates with 15-minute exposures. Special areas of the southern sky were also mapped spectroscopically with the UK Schmidt using objective prisms.

In 1964, the KSO Schmidt began work on the Tautenburg Astrophysical-Statistics Survey of Galactic and Extragalactic Objects. The fields included in this plan were chosen according to very strict criteria:

- The fields should contain faint standard stars, so that magnitude scales in the UBV (Ultraviolet, Blue, Visual) system could be established to low limiting magnitudes;
- The fields should have low, well-established, interstellar extinction;
- The fields should be such that they would allow significant characteristics of the distribution of galaxies throughout the universe to be determined.

A total of 300 square degrees met these and other criteria, were included in the programme, and were photographed in 5 colours (Ultraviolet, Blue, Visual, Red, and Infrared).

Another extensive programme with the KSO Schmidt was designed to establish a set of 'reference points' in the sky, formed of compact galaxies without any proper motion. In the first phase of this programme a complete zone at declination of 52.5° was photographed and analysed.

Apart from such survey programmes, large Schmidt telescopes are of course also used for the specific investigations of interesting individual objects. Needless to say, the ESO and UK Schmidt telescopes, for example, were used to examine the Magellanic Clouds complex, and the KSO Schmidt was similarly used on the Andromeda Nebula. In addition, questions often arise from the survey work, and these have to be looked at in more detail.

The great variety of results from the survey programmes and other observations with Schmidt telescopes are far too numerous to be described here in detail. We can say, however, that Schmidt observations have been, and still remain, of great significance in nearly all areas of astronomy, and this may be seen from a few examples. Examination of galactic structure is a classic field of Schmidt astronomy. Observational data from Schmidt telescopes have made decisive contributions to the examination of the spatial distribution of the various populations in the galaxy; defining the position of the spiral arms from young, hot stars, and determining the shape and extent of the galactic halo from old stars. A fundamental problem of galactic astronomy is the establishment of the luminosity function, above all, in the region of the faintest stars. These results are also mainly based on Schmidt data.

Spectral surveys with objective prisms have led to the discovery of numerous groups of very interesting stars. These include the Wolf-Rayet stars, peculiar-A stars, Hα-emission stars, etc.

Of particular interest in the study of the evolution of the Galaxy are investigations of the first generation of stars in the Galaxy, i.e. of metal-poor stars. The answer to this question is yet another that has to be obtained largely from observations by Schmidt telescopes.

The investigation of interstellar matter is of special significance for explaining the evolution of the Galaxy, because the interstellar medium is, on the one hand, the material from which stars are born, and, on the other, is enriched with heavy elements and dying stars, for example, through supernova processes. Interstellar material often appears in the form of extended objects, so Schmidt observations are particularly suitable for their examination.

Observations with Schmidt telescopes are also very successful in extragalactic astronomy. The examination of the nearest galaxies (the Andromeda Galaxy and the Magellanic Clouds) has already been mentioned. In addition, we should also mention the compilation of comprehensive catalogues of bright and faint galaxies, and of clusters of galaxies. Quasars were originally discovered through the identification of radio sources. Later, however, the number of known quasars increased quite dramatically, because of photographic and photometric observations with Schmidt telescopes. In addition, many quasars were found by the detection of their strong emission lines on objective prism plates taken with Schmidt cameras.

The photographic plate as a detector medium

The successes mentioned are only a selection and it should be emphasized that the use of Schmidt telescopes has had a major influence on practically all fields of astronomy.

The outstanding advantage of Schmidt systems lies in their large field of view, which demands large detector areas. With the large Schmidt telescopes these measure 24 cm × 24 cm and 36 cm × 36 cm, but, even in the case of Schmidt systems with 60-cm corrector plates, image surface dimensions are 16 cm × 16 cm. In the last 25 years, electronics has undergone a particularly major evolution. Numerous electronic detectors have been developed and these are also used in astronomy. This development has often given rise to the opinion that photography would no longer have any role to play in astronomy. We should perhaps bear in mind here the anecdote about Mark Twain. On reading his obituary in a newspaper he remarked that rumours of his death were greatly exaggerated. Mark Twain lived for a long time after the appearance of his premature obituary, and this is also true of astronomical photography. Periodically, one hears that photography in astronomy is finally going to be replaced by electronic methods. Despite this, the significance of photography has stubbornly continued to grow, especially with the commissioning of new telescopes with large fields of view, such as Schmidt telescopes. Some of the reasons for this are as follows:

- As yet, electronic detectors can only be produced with maximum dimensions of 5 cm × 5 cm, i.e. 25 cm². But photographic plates can easily be obtained up to 40 cm × 40 cm. If the whole of the coma-free and vignetting-free fields of Schmidt telescopes are to be used, we are forced to employ photographic plates, because even moderate-sized Schmidt telescopes have image surfaces that cover some 250 cm² – ten times that offered by electronic detectors. Large Schmidt telescopes even have image surfaces of between 600 and 1300 cm².
- Further advantages of photographic plates are their high information content and the simplicity with which they may be handled and analysed. A typical fine-grain emulsion has an information content of one million bits per square centimetre.
- It should also be mentioned that, in comparison with electronic detectors, the photographic plate is inexpensive and electronic detectors need additional electronic equipment and storage media.

For all these reasons, but particularly because of its large surface area, the photographic plate will remain the most important detector for use with Schmidt telescopes for a long time to come.

The photographic plate also has its disadvantages. The emulsion, for instance, has a quantum efficiency of only about 1%. The cause of this is that only 10% of the photons falling on the emulsion layer are absorbed by the light-sensitive silver bromide crystals. A portion of the photons is reflected by the layer, and another fraction enters the emulsion without encountering silver bromide grains, which are suspended in a layer of gelatine. A silver bromide grain can be developed – i.e. changed into black metallic silver – only after it has been hit by not just one, but by about ten, photons, which it absorbs. It should be noted that silver bromide grains that absorb less than ten photons are not capable of being developed. These photons are lost from the actual image-producing process, because the grain of silver bromide hit by them does not contribute to any blackening of the plate. The result of this is that only about 1% of the photons actually contribute towards the formation of an image.

In contrast, electronic detectors have considerably higher quantum efficiencies and easily achieve a quantum efficiency 50 times that of a photographic emulsion.

As already mentioned, the grains of silver bromide that have absorbed sufficient photons are reduced to black metallic silver during the development process. The unexposed grains of silver bromide are removed from the gelatine layer by the fixing bath. There is therefore a connection between the number of photons incident on the layer, i.e. between the amount of light and the resultant degree of darkening. The number of incident photons is determined by the intensity (I) of the source of radiation, and by the length of time (t) during which photons encounter the layer.

The connection between the number of incident photons and the resulting density is known as the characteristic curve, and unfortunately this is not linear: which is a further disadvantage of photographic emulsions.

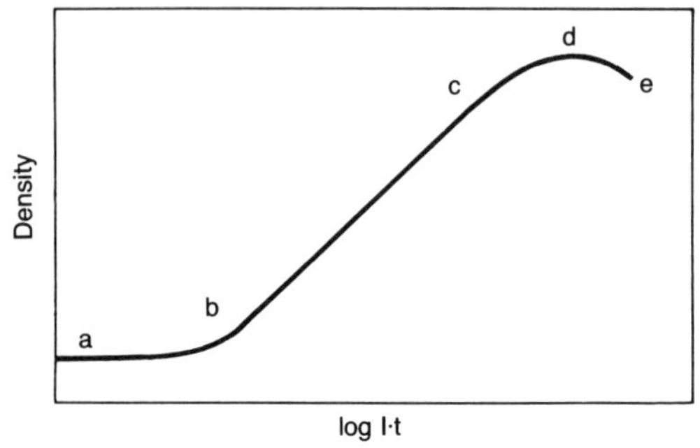

Fig. 4.2. Four different parts of the characteristic curve are indicated on this diagram. The slope of the linear portion of the characteristic curve shows the contrast. The range of intensities that can be covered is dependent on the slope of this linear portion. The flatter the curve, the greater the range that can be recorded. (a–b: the underexposure region; b–c: straight-line portion; c–d: over-exposure region; d–e: solarization.)

It is easy to understand that a certain minimum number of photons is required in order to cause any blackening. This minimum value that must first be attained is known as the threshold (Fig. 4.2). If the number of photons then increases, the density slowly and gradually becomes greater. The increase in density eventually becomes constant. This indicates that one has reached the linear part of the characteristic curve. With any further increase in the number of photons, the density increases more slowly, and the slope of the characteristic curve becomes flatter.

In fact, the density produced is not simply dependent on the product of intensity and time ($I{\cdot}t$), but – as has been shown in practice – varies as $I{\cdot}t^P$, where P is the Schwarzschild exponent (named after Karl Schwarzschild) and which, in the case of good astronomical plates lies between 0.8 and 0.9. The effect of the Schwarzschild exponent is that every increase in exposure time has a decreasing effect on the image density. This can be seen quite clearly in Table 4.1. If $P = 0.8$, then an exposure increased by a multiple of five only produces 3.6 times the effect; an exposure increased 10 times has 6.3 times the effect; and an exposure increased by 20 times has just 11 times the effect on the resulting density.

The shape of the characteristic curve and the existence of the Schwarzschild exponent make it more difficult to work with the photographic plates.

The grains of silver are able to absorb photons corresponding to wavelengths between 250 and 550 nanometres – i.e. photographic emulsions are sensitive to radiation over this range of wavelengths. There is, however, other longer wavelength radiation from the stars that we would also like to detect with Schmidt telescopes. For this, photographic emulsions have to be specially prepared. Dyes are included in the layer of gelatine that absorb radiation at long wavelengths and transfer the energy to the grains of silver

Fig. 4.3. The curves show the sensitivity versus wavelength for typical astronomical photographic plates (Orwo ZU 21; Kodak 103 a–D; Kodak 103 a–E). These emulsions can cover the total range of optical wavelength from 300 nm to 700 nm.

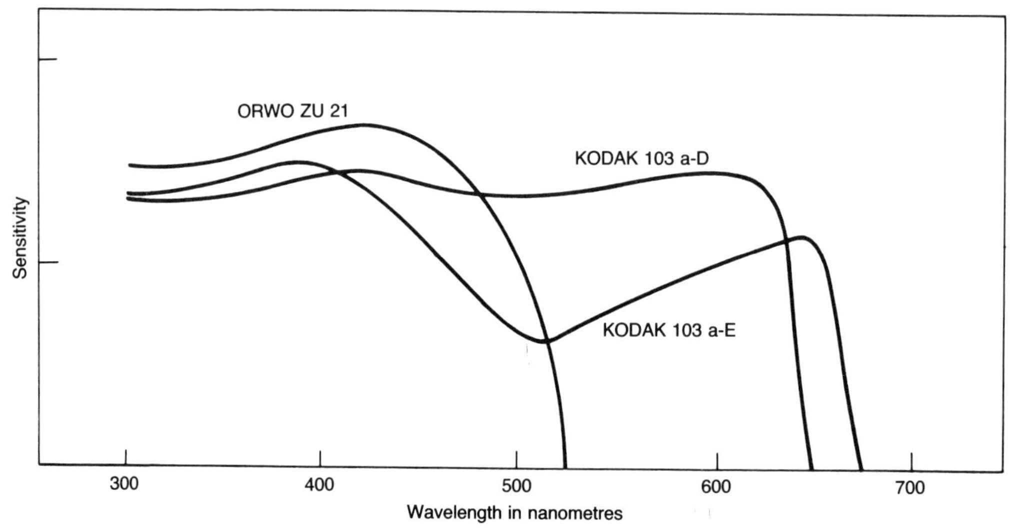

Table 4.1. *The relationship between exposure time, intensity and the Schwarzschild exponent* ($P = 0.8$)

Exposure time, t	1×	5×	10×	20×
t^P	1.0	3.6	6.3	11.0

bromide. According to the dyes used in the layer of gelatine, one can obtain emulsions that are sensitive to red or even to infrared radiation. Some examples of photographic plates that are sensitive to different wavelengths are shown in Figure 4.3.

If stars are photographed with the Orwo ZU-21 plate or the Kodak 103a-O plate, only the short wavelength, blue part of the radiation is recorded; the long wavelength, red radiation is not registered. If the red part of the spectrum is to be recorded, then a Kodak 103a-E plate, for example, has to be chosen for the observation. As can be seen quite clearly from the curve showing sensitivity versus wavelength, the Kodak 103a-E records both the red and the blue regions. In order to avoid this, a filter that transmits only long wavelength, red radiation has to be placed in front of the photographic plate (Fig. 4.4). By careful combination of plates and filters in this way, various wavelength regions can be achieved. In order to gain a first

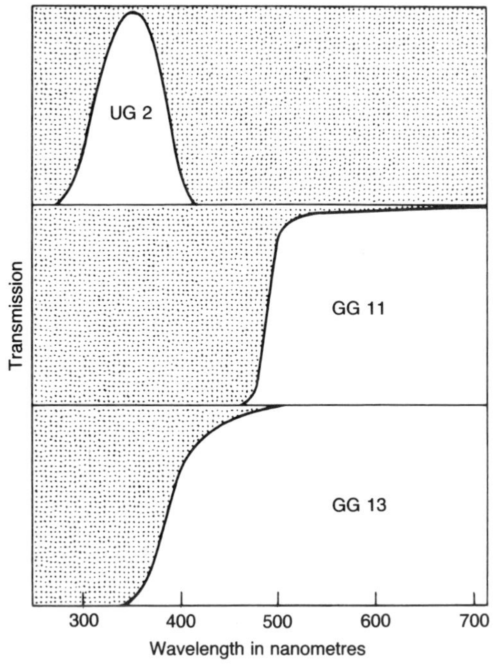

Fig. 4.4. This diagram shows the transmission of three filters that are frequently used in astronomy. The UG 2 is used for the exposures in the ultraviolet U band. It transmits radiation between 300 and 400 nm and has its maximum transmission at 350 nm. The transmission of the GG 13 filter begins at 375 nm, and that of the GG 11 not until 475 nm. The actual transmission also depends on the thickness of the filters.

impression, the sensitivity curve of a photographic plate and the transmission curve of a filter may be superimposed, and one can see immediately over which range of wavelengths the combination could be used for observation (Fig. 4.5).

An important advantage of a photographic emulsion is that it integrates the incident radiation – i.e. the longer photons from a source of radiation affect the photographic plate, then the greater the resulting density, until solarization is reached. From this it might be concluded that faint stars would become increasingly easy to detect the longer the exposure. But photons do not come from just discrete sources of radiation – i.e. stars – but also from the overall sky background. This means, however, that the whole of the photographic plate is affected by the sky background. The density depends on the brightness of the sky background. This means that, with Schmidt telescopes, the optimum exposure time is one that produces, for the given sky brightness, a background density of 0.6 to 0.8. The sky near the Moon, for example, is very bright, and, during observation with Schmidt telescopes of such areas of the sky, the general plate-density reaches very high levels after just a short period of observation. This example shows that, with Schmidt telescopes, exposure times are not unlimited, because the plate would then be too dense for further processing. The sky brightness limits the exposure times.

Fig. 4.5. Combining a KODAK 103a–D photographic plate with a GG 11 filter approximately matches the visible spectral response. The eye's maximum sensitivity lies at about 550 nm, and this is closely matched by this emulsion-filter combination. Radiation below 475 nm is not transmitted by the filter and the emulsion is not sensitive to radiation above 650 nm. In order to calculate the exact range of wavelengths covered, the shape of the transmission curve, the filter thickness, and the exact sensitivity curve of the emulsion all have to be taken into consideration. With Schmidt telescopes the spectral transmission of the correction plate also plays a part, as well as reflectivity of the mirror at different wavelengths.

There is one other point to be considered in the determining of exposure times for observations with Schmidt telescopes. The photons that come from a star are the signal photons, those that come from the background cause noise, which is interference. As the signal-to-noise ratio should be as great as possible, the ratio between the signal and the noise determines the accuracy with which objects on photographs can be measured photometrically. Experience has shown that the most favourable signal-to-noise ratio is obtained when the background density is between 0.6 and 0.8. This means that, with Schmidt telescopes, the optimum exposure time is one that produces, for a given sky brightness, a background density of 0.6 to 0.8.

There are various causes of the fluctuations in the brightness of the sky. One obvious variation in the sky brightness is that caused by the different phases of the Moon – Apart from this, local atmospheric conditions contribute to variations in the sky brightness. A high proportion of dust in the atmosphere scatters light over large areas of the sky thus brightening it. Artificial-light sources on Earth also brighten the night sky. Optical observations cannot be made with Schmidt tele-

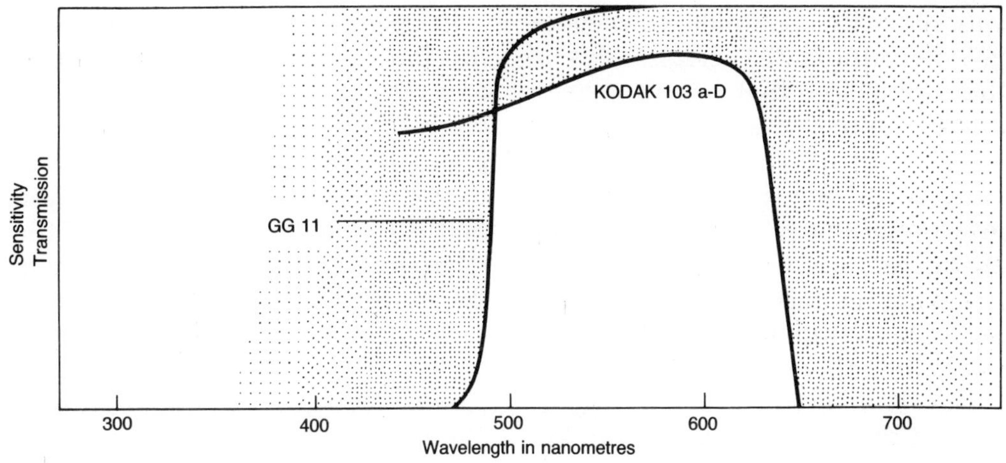

scopes at the time of Full Moon, and large conurbations with a lot of artificial-light sources close to the observatory are important sources of light pollution.

As the sky brightness determines the optimum exposure time for photographic observations with Schmidt telescopes, it is advantageous if it can be measured objectively. The measurements should, of course, be made over the spectral regions, in which photographic observations are being carried out. On the Schmidt telescope at Tautenburg, the brightness of the sky is measured with a photoelectric photometer, and therefore the optimum exposure time may be determined for each observation. This photometer also works in the UBV-system. Measurements of sky brightness are carried out during the photographic exposure on the field that is being recorded by the Schmidt telescope. Care has to be taken that the photoelectric photometer's field, which has a diameter of 1°, does not include any bright stars that could distort the measurement. For this reason the photometer's alignment can be altered slightly.

Reduction and analysis of photographic plates

When observations are made with Schmidt telescopes, the photographic plate is the detector and it also stores the information which can then be analysed at a later date. The methods of analysis have to be suitable for the high information content and the large format of the photographic plate. Assuming that 10^6 bits are stored in every square centimetre, then a $24\ \text{cm} \times 24\ \text{cm}$ plate contains about 6×10^8 bits, and a $36\ \text{cm} \times 36\ \text{cm}$ plate holds 1.3×10^9 bits. One of the most important aims of making observations with Schmidt telescopes is determining the magnitudes of the objects that have been photographed in the various spectral regions. These magnitudes then form the basis for further interpretations and scientific advances.

Taking the formation of an image on a photographic plate as a starting point, the density of a stellar image is a measure of the brightness of the source of radiation. We know, however, that within the emulsion the photons are partially scattered away from the true image position. This causes the sizes of even the faintest stellar images to be enlarged. The diffusion effect is greater the more photons enter the emulsion. This means that the final image diameter is also a measure of the brightness of the star. So it is essential to find a method that will simultaneously take the density and the diameter of a stellar image into account as a function of the brightness of the star.

In order to measure the density, the dark stellar image is illuminated by light from a photometer lamp of known intensity (I), and a photoelectric photometer is used to measure the intensity (i) of the light that is transmitted. The ratio of the intensities (I/i) is then a measure of the density, and may be expressed as $D = \log I/i$. So we have a density of 1, if only one-tenth of the light passes through the photographic plate, a density of 2 when one-hundredth passes through, and so on.

We can now interpose a diaphragm in the light path, and vary its diameter, so that, irrespective of the density of the stellar image and its diameter, the same amount of light passes through the plate all the time. The diameter of the diaphragm opening, which is determined by the density and diameter of the stellar image is thus a measure of the magnitude of the star. Such a variable diameter diaphragm is known as an iris diaphragm, so this special instrument for measuring the brightness of stars is known as an iris-diaphragm photometer. With such a measuring procedure, errors can arise because of fluctuations in the brightness of the lamp (Fig. 4.6). This can be determined by passing some of the light on a separate light-path into the photometer and then alternately measuring the amount of light passing through the photographic plate, and through the control light-path.

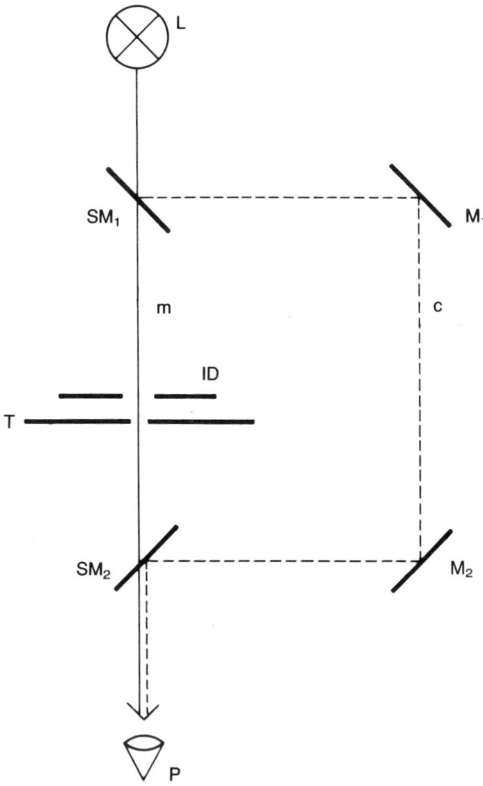

Fig. 4.6. This diagram shows the principle of a double-path, iris-diaphragm photometer. In practice, the size of the iris diaphragm (ID) is adjusted when measuring different stars so that the amount of light (m) passing through the plate is the same as the amount in the control light-path (c). (L – photometer lamp; SM – semi-silvered mirror; M – mirror; T – plate table; P – measuring device)

The iris-diaphragm photometer serves to measure individual points, i.e. individual stars on the plate (Fig. 4.7). Another possibility is to record all the information on the plate and store it in a large computer or in an electronic database. In order to do this, the photographic plate has to be scanned in lines and the density measured point by point.

Finally, interesting information may be extracted from the mass of data that has been obtained by the image processing. We can, for example:

- obtain, from the computer, the positions of all the stars differing greatly in brightness in two different spectral regions,
- have the computer print out all the stars that are particularly bright in one spectral region, e.g. in the blue,
- indicate the stars that have different magnitudes on two plates, i.e. that are variable stars,
- investigate galaxies, etc.

Examples of universal measuring instruments of this type that are used to analyse Schmidt telescope observations are the COSMOS machine at the Royal Observatory, Edinburgh, and the GALAXY machine at Cambridge Observatory.

In general, it may be said that, when using Schmidt telescopes, the time spent analysing the photographs is much longer than the time taken by the observations. The total number of all the plates taken with the six largest Schmidt telescopes so far certainly amounts to more than 100,000. The proportion of the information that they contain and that has already been reduced and analysed is difficult to estimate; it is certainly under 20% and may even be below 10%. This becomes clear if we consider the exposures that were obtained of just one special object – for example, a comet. The area occupied by the head of the comet that has to be examined amounts to several thousandths of the total area of the photographic plate (Fig. 4.9), but, apart from the comet, there are many tens of thousands of stars and galaxies on the plate, which could also be the subject of specific investigations. The same holds for exposures that were made for the examination of specific star clusters, clusters of galaxies, planetary nebulae, interstellar dust clouds, etc.

Photographic plates made by Schmidt telescopes are frequently analysed several times for different purposes. At the Karl Schwarzschild Observatory all exposures, irrespective of the scientific programme for which they were made, are routinely scanned for minor planets, although minor-planet research is not one of the fields of work carried out at KSO. It is, however, possible to provide specialists in this field with between 500 and 700 positions of minor planets annually, and, in the process, several new minor planets have been discovered.

These simple examples show the great variety of ways in which the observational material obtained

Fig. 4.7. The iris-diaphragm photometer at the KSO, Tautenburg was developed at the observatory. In photometric work it is, in practice, necessary to measure several photographic plates. Generally plates taken in different spectral bands are involved, and several plates are evaluated in each wavelength region in order to reduce any errors of measurement. The coordinates of the points measured on the first plate of a series are therefore stored by the computer in this iris photometer, and the subsequent position of the photometer's plate-holder is controlled automatically by the computer when measuring the other plates. The actual measurement procedure, i.e., comparison of the intensity of the light in the measurement path and in the control path also takes place automatically. The aperture of the iris diaphragm is taken as a measure of the brightness, and is stored by the computer for further processing.

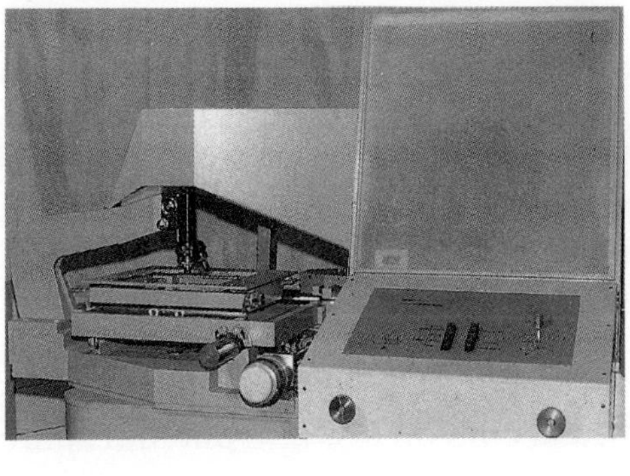

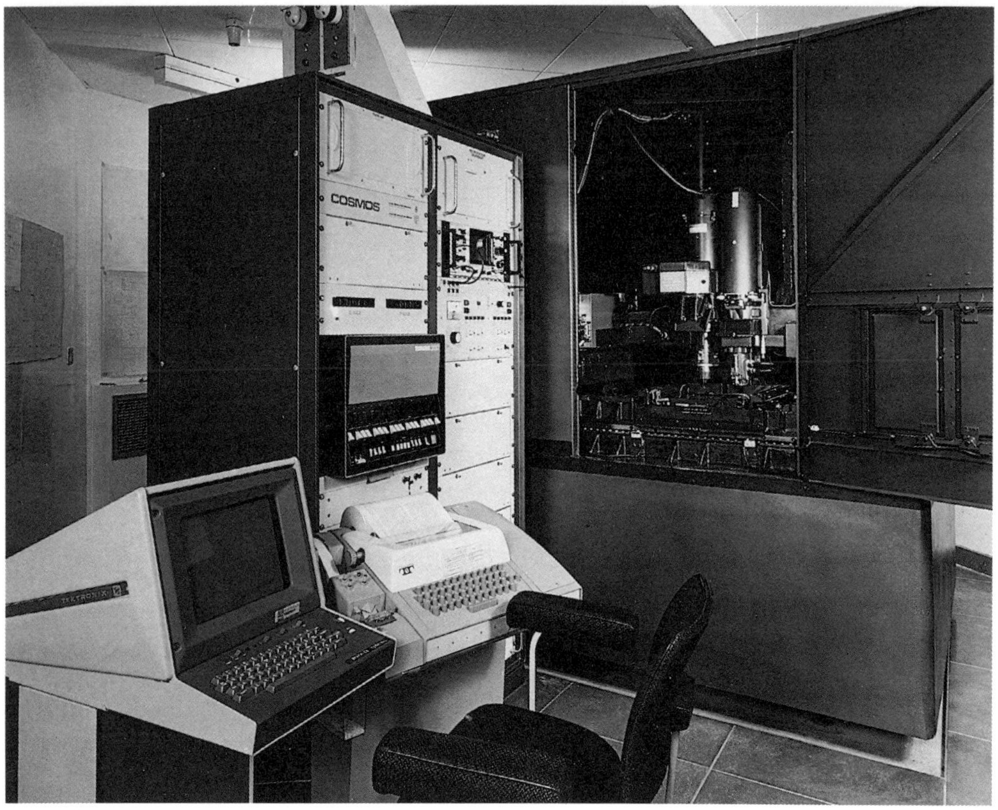

Fig. 4.8. The COSMOS high-speed measuring machine at the Royal Observatory, Edinburgh, serves to obtain information recorded on photographic plates. The basic mechanical and optical components of the machine can be seen in the right-hand half of the picture.

The photographic plate may be up to 356 mm square. The plate, on its carrier, may be positioned to an accuracy of 0.5 µm on both the x- and y-axes. The plate is scanned with a narrow beam of light and the amount of light that passes through the plate is a measure of the density at each point. The two racks in the centre of the picture contain the electronics, including a microcomputer, which controls the operation of the equipment. The operator can enter the appropriate commands from the keyboard on the left. An immediate printout of the data is available if required.

The machine can scan an area 200 mm × 200 mm, with a resolution of 8 µm, and analyse the images in 10 hours. The machine can store the data from 400,000 image points on tape.

with Schmidt telescopes may be examined. There is also often considerable interest in making copies on photographic paper for visual examination.

Astronomical photographic emulsions are generally very high in contrast. This means that a slight increase in the intensity of the illumination leads to a significant increase in the density. When photographing many extended objects such as the Orion Nebula, for example, a wide range of densities occurs on a single photographic plate. There are areas which are very low in density and that transmit nearly 100% of the light, and other areas have a density of around 4 and only transmit 0.01% of the light. The transmission ratio of 1 : 10,000 in the original cannot be reproduced on photographic paper. The range of densities that can be reproduced is around 1 : 30. A lot of the information available in the original negative is lost in reproduction. In order to record photographically all the information in an object with higher density ratios, it would be necessary to produce a whole series of exposures with different exposure times, which for a transmission ratio of 1 : 10,000 would lie approximately between 5 seconds and one hour (Fig. 4.10). Each individual copy would then record only one particular part of the large density range.

Because of these problems in photographic reproduction, various methods have been developed to retain a greater information con-

Fig. 4.9. This plate of Comet IRAS-Araki-Alcock was made on May 9 1983, with the 2-m Schmidt telescope at the KSO, Tautenburg. The exposure was ten minutes on Orwo ZU 21 with a GG 13 filter. It indicates very clearly how the comet, on which the exposure was guided, only occupies a very small area of the plate. Because it had a very high proper motion, all the other objects appear as linear trails.

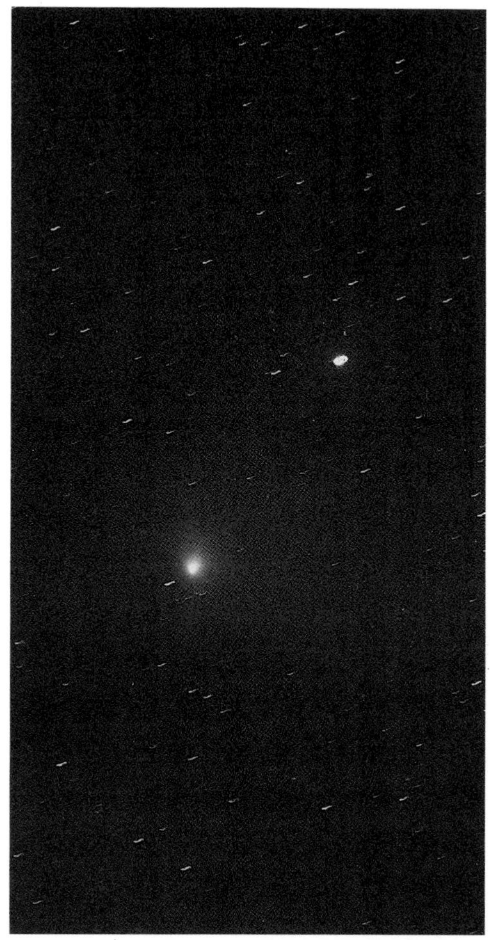

tent. David Malin's unsharp masking technique is one that is frequently used. An unsharp mask is produced by making a positive contact print from the original negative onto a low-contrast film. As the film is brought into contact with the back of the original plate, and a diffused light-source is used, one gets an unsharp copy. If the unsharp positive is then placed in contact with the original negative most of the large-scale structure is masked, so only the fine details remain when a picture is made using the original negative and the unsharp mask. An impressive example of this is shown by the picture of the Orion Nebula (Fig. 4.11).

Fig. 4.10. These copy images of the Orion Nebula were taken from a single negative using varying exposure times. They clearly show how only a very small range of brightness can be reproduced at any one time. The extended nebulosity may therefore either be clearly visible, or be completely invisible, but with the Trapezium stars appearing in its place.

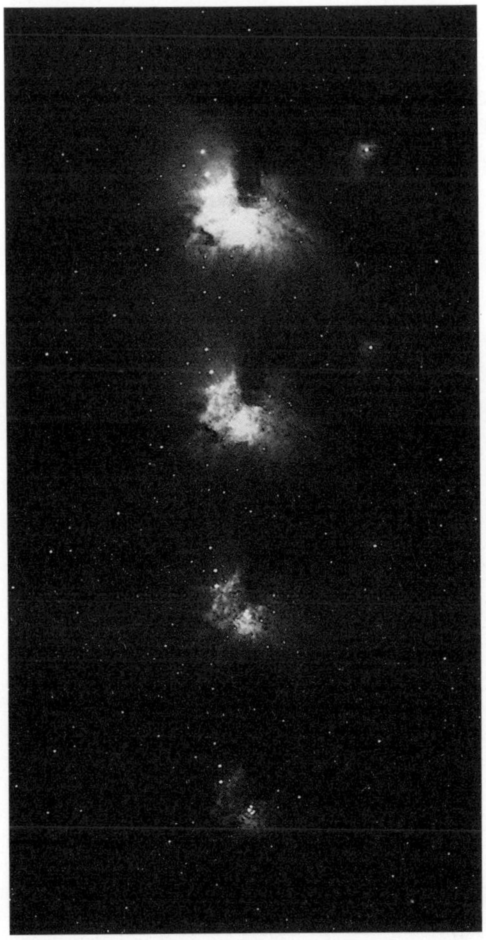

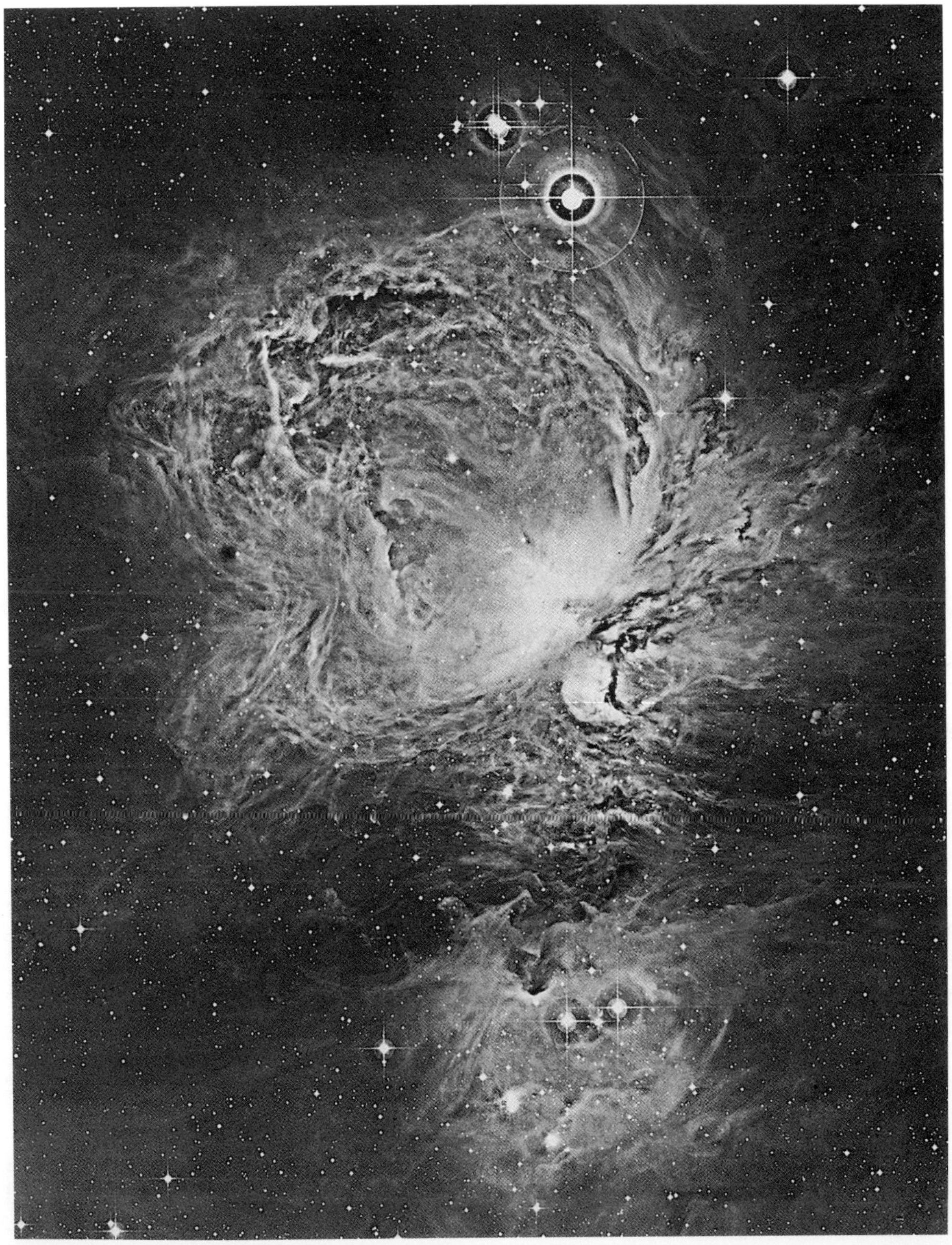

Fig. 4.11. This illustration shows the Orion Nebula after the image has been manipulated using the unsharp-masking technique. With this method the Trapezium stars in the brightest area of the Nebula are visible, but the structures within the nebulosity itself is also retained. A large range of differing density on the original plate is therefore reproduced.

Another process developed at the KSO in Tautenburg was used to produce the pictures for the illustrations in this book. In this process, a direct contact copy is produced from the original negative onto high-contrast film with 1000 times over-exposure. Before development, special treatment causes halogens to be recaptured by the nuclei on the surface of the grains in the grossly over-exposed latent image. However, the metallic silver image in the very centre of the grains is not affected by this process. The latent image is then developed. This method enables copy transparencies with a great tonal range to be obtained, which attain nearly the greatest densities present in the original while at the same time showing no diffusion or loss of detail in the grossly over-exposed areas. The photographic reproductions of the objects completely retain their natural appearance with this process.

Amateur astronomy with Schmidt telescopes

It is not surprising that the Schmidt-telescope design, with its special properties, is exceptionally important to professional astronomers. Quite apart from that, however, it is also a very attractive optical system for those who actively carry out serious astronomy in their spare time – i.e. the extremely numerous amateur astronomers found all round the world. Some of them are satisfied with the pleasure of expanding their own knowledge of the sky, or in gaining additional insight into astronomical events. But many of them are also seriously concerned with research, either aiming to discover, or to follow, specific astronomical phenomena. They frequently use instrumentation specifically designed for the task, such as telescopes and accessories that they have taken great pains to construct themselves. The work that they undertake can include anything: from grinding and polishing the optics, machining the mechanical parts, and constructing shelters for the telescopes, to linking in electronics and computers.

What relevance do Schmidt telescopes have for active amateur astronomers?

Any amateur astronomers who take their hobby seriously try to develop their observational techniques in a manner appropriate to their specific interests. As already mentioned, many of them derive great aesthetic pleasure from some of the beautiful objects in the sky – not just in observing them visually, but also in obtaining photographs; the latter can be examined at any time and can be shown to friends and acquaintances.

The most attractive objects in the sky are, however, extended objects such as luminous interstellar nebulae, nearby galaxies, globular clusters, etc. Objects with a large surface area require instruments with a large field of view, but – and this is of particular importance – the diameter of the free aperture of the telescope does not play a part.

The brightness of the image of a point source is proportional to the square of the diameter of the objective, i.e. the larger the aperture, the fainter the point sources that are detectable. For extended objects, however, the brightness is proportional to the square of the focal ratio (the focal ratio being the diameter of the free aperture divided by the focal length). When the aperture is increased, more radiation is captured, but, if the focal length increases at the same time, the radiation is spread over a greater detector area. A small Schmidt telescope of 300-mm aperture is just as successful at imaging extended objects as the largest Schmidt telescope on Earth (with an aperture of 1340 mm) if it has a focal length of 900 mm instead of 4 m. In both cases the focal ratio is *f*/3. However, amateurs have to remember that their images are smaller. This relationship is strikingly shown if we compare two images of the North America nebula, one taken with the original Schmidt camera, and one with the KSO Schmidt at Tautenburg (see pp. 141 & 143). Depicting extended objects using small-aperture Schmidt telescopes with correspondingly short focal lengths is therefore a very worthwhile task for amateur astronomers.

Important, and successful, scientific work can be undertaken by amateurs thus equipped, by detecting objects whose outbursts cannot be predicted, and by monitoring objects that vary in brightness. The more observers with wide-field telescopes, the greater the chances of success. In general, there is very little time available for large Schmidt telescopes to follow such monitoring programmes.

Observations undertaken for scientific work should not be made visually, but with objective radiation detectors. From this point of view, the photographic plate is very suitable for amateurs, because it simultaneously serves as a detector and as a storage medium, is very cheap, and is also simple to use. Standard Schmidt telescopes can only be used photographically, so in this respect they appear very 'user-friendly'.

Many amateurs are not satisfied with just collecting photographic plates, but work with a specific aim in mind. This may be the measurement of exact positions or determining the brightness of specific objects. There is a whole series of different procedures that can be used for photometry alone, and these often involve a considerable amount of instrumentation and effort. The apparent brightness of objects, for example, is dependent on the diameter of the stellar images, among other factors. A simple method of determining stellar magnitudes therefore consists of measuring the diameter of the images. Here one of the 'disadvantages' of the Schmidt system becomes apparent: the images are often too sharply defined. Determining brightness from the diameters therefore becomes very difficult, and a more complicated method of measurement has to be employed.

Since 1926, there has been an annual gathering of American amateur telescope-makers near Springfield in Vermont. On an open-air ground that at times resembles a pop festival, they display their equipment, gaze in admiration at one another's work, and become involved in detailed discussions of the instruments. The 'Stellafane' meetings are also used to display telescopes of unusual optical and mechanical design and construction. Over the years, this exhibition – which is certainly the largest in the world – has shown a representative selection of the types of telescopes constructed by amateur (or at least American amateur) telescope-makers. True Schmidt telescopes are rarely seen, however. There are nevertheless many other aplanatic reflecting systems, which are able to provide a larger, coma-free field through special optical design.

The preference for special types obviously springs from the fact that amateurs are not solely interested in photographing extended areas of the sky, but that they are particularly concerned to use reflectors for the visual observation of objects. Undoubtedly they are also influenced by the desire to find other solutions to the optical problems and to exhibit their technical skills. The latter are required in manufacture, if the one disadvantage of the Schmidt system, its great overall length, is to be overcome by modified optical configurations. Under these circumstances, apart from the corrector plate, other aspheric surfaces may have to be manufactured.

A frequently chosen variant is the Schmidt–Cassegrain system. It differs from the original version devised by Bernhard Schmidt in that a convex, aspheric mirror is fixed to the rear surface of the corrector plate, and reflects the beam of light from the primary mirror back into its original direction. If the overall focal length thus obtained is made sufficiently long, and if the primary mirror is perforated in the centre or a small plane mirror mounted in a Newtonian-type arrangement, the light may be brought outside the tube, where it can be examined visually with an appropriate eyepiece. This solution offers both the advantages of the Schmidt form, i.e. a large field of exceptional image quality when compared with conventional Cassegrain systems – and of shortening the overall length to approximately one half. That also greatly increases the ease with which the telescope can be handled: it is easy to switch from one object to another when undertaking visual work. Transporting the instrument to the observing site is also considerably simplified because of the smaller overall size. However, a greatly shortened focal length has to be paid for in the use of an aspheric primary mirror. In the Schmidt–Cassegrain system, the diameter of the corrector plate can be almost as large as that of the primary mirror, so the latter's free aperture is fully utilized.

The manufacture of the corrector plate – which has a stronger curve than that of the original Schmidt design – needs special optical techniques and considerable effort. Telescope makers use either Schmidt's original techniques of deforming the glass plate with a vacuum when grinding and polishing a spherical, or even a plane, surface; or else they try to obtain the required figure by selective grinding and polishing. In the latter case, appropriate optical tests have to be applied throughout manufacture in order to determine which zones require further work. The total thickness of glass to be removed in making a Schmidt plate is very small, being of the order of ten wavelengths of light, i.e. a few hundredths of a millimetre.

A reduction in the overall length to approximately the same as the focal length was achieved in 1935, in a system described by F. B. Wright, which also provided a flat image plane. This was accomplished solely by modifying the form of the two optical elements, the primary mirror and the corrector plate, and considerably simplified the design's use for visual work. Here, the surface curve on the primary mirror was an ellipse, whose semi-major axis coincided with the optical axis.

Another aplanatic system, more frequently constructed by amateurs, is the Maksutov system. Here, the spherical aberration of the primary mirror is corrected by a strongly curved, meniscus lens with two spherical surfaces. If a particularly fast system is required, with a large aperture relative to the focal length, deviations from spherical surfaces become necessary, making construction much more difficult.

Bernhard Schmidt himself mentioned the possibility of dispensing with the corrector plate altogether, with a view to making the design technically simpler. The system then reduced to a spherical primary mirror and a smaller aperture stop, placed at the centre of curvature. However, correction for the spherical aberration of the primary mirror is lacking, so the image quality is naturally degraded. The faster the system, the worse the quality becomes. This raises the question of what degradation can be tolerated. Schmidt himself suggested a limit of eight times the diameter of the aperture stop (i.e. an *f*/8 system); practical experiments by amateur astronomers have given good results even down as far as *f*/5. The basis of this concession to poorer image quality is that, in practice, even a completely aberration-free photographic system has to be used through a turbulent atmosphere, and therefore can never give the theoretically attainable performance.

As an example, let us take an amateur Schmidt telescope with an aperture stop 150 mm in diameter, and a focal length of 750 mm (*f*/5). Apart from optical errors, the resolution theoretically attainable will be determined by the unavoidable diffraction at the aperture stop. With the sizes given this amounts to one second of arc, and corresponds to exactly 4 μm on the photographic plate. Changes in direction caused by scintillation, itself produced by turbulence in the Earth's atmosphere, caused the images to be spread out to an average of two or three times this amount. In addition, the resolution of a photographic emulsion is about 20 μm. As a result, it is not even possible to approach the theoretical resolution of the optical components of the system.

The linear diameter of the bundle of rays forming a faint stellar image at the image surface in a correctorless Schmidt system, as derived from theoretical calculations or ray-tracing on a computer, is

$$d = 0.008\,\frac{D^3}{f^2}$$

where D is the free diameter of the aperture stop and f is the focal length. Both quantities are to use the same unit (e.g. millimetres). For the example mentioned above, this equation gives an image diameter of about 50 μm. If that seems too large, a slight increase in the focal length will produce a considerably better figure. Even without a corrector plate, the system remains coma-free, i.e. the star images are symmetrical to the very edges of the wide field. Despite its simplicity, this combination of a spherical primary mirror and an aperture stop is therefore superior to many other solutions.

The advantages for photometric work using simple methods, such as measuring image diameters, are obvious.

Although for most professional scientific purposes the Schmidt telescope is constructed in its original form, certain specific other types have their uses for amateur astronomers. Even choosing the most appropriate variant for the purpose in mind is an interesting and worthwhile task.

5 The life and work of Bernhard Schmidt

The name of Bernhard Schmidt is known today to astronomers all over the world, and all are aware of the significance of the optical system that he developed as well as the research results that it has produced. Bernhard Schmidt (Fig. 5.1) did not live to see the great success of his telescope design, because this occurred nearly two decades after his death. From the point of view of astronomy, people mainly consider the invention of the coma-free mirror as Schmidt's greatest achievement, but he was also involved with many other technical problems. Bernard Schmidt's independent, and often unconventional, preoccupation with technical questions arose unprompted by anyone else and started when he was young, although there was little experience or example of this in his parents' household.

Bernhard Schmidt's father, Karl Konstantin Schmidt, was parish clerk of Naissar, which consists of a small island in the bay of Tallinn, in Estonia. Naissar itself was administered from Tallinn. The parish was only small: in 1816 it registered 97 inhabitants; in 1844, 139; in 1856, 200. Documents show that between 60 and 70% of the inhabitants of Naissar were Swedes.

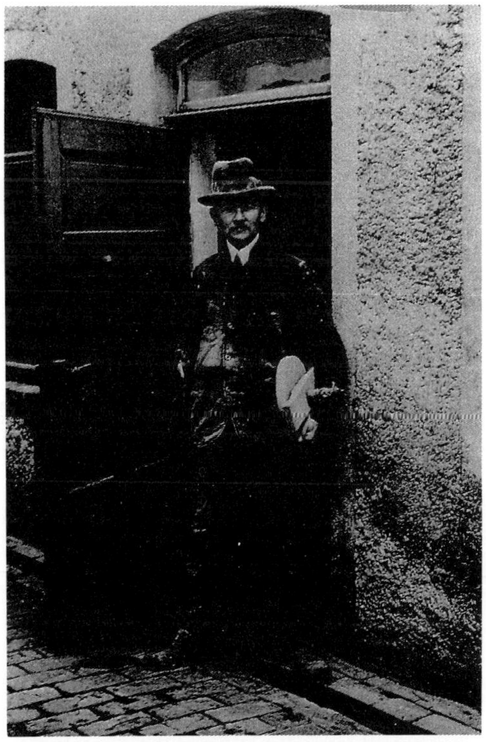

Fig. 5.1. This photograph shows Bernhard Schmidt outside the door of his workshop in Mittweida.

As well as being parish clerk, Bernhard Schmidt's father also farmed 5.4 hectares of land, which he had taken over from his father along with the house, and he also spent some time fishing. This indicates that the salary of a parish clerk was not sufficient to support a family and that farming and fishing were therefore necessary as additional sources of income.

The inhabitants of a small island quite naturally turn to fishing and for most people this was the main occupation. The islanders also described themselves as pilots, although undoubtedly they only rarely carried out such work. This is, however, probably the reason why Bernhard Schmidt always gave his father's occupation as pilot.

Bernhard Schmidt's mother was called Maria Helene Christine and her maiden name was Rosen. Her father, too, was a pilot on the island of Naissar.

German family names were not uncommon in Estonia in the last century. It should not be inferred from this, however, that the family was also of German ancestry. For some time in Estonia the Germans formed an upper class which also included the priests. The latter were responsible for all the details recorded in the church registry. Those who could not give a surname for an entry, were given one by the priest. Many families were given German names, and this is probably how the Schmidt family name got into the church records of Tallinn.

Childhood and youth

Bernhard Schmidt was born on Naissar on 11 April 1879, the first of five children. Little is known about his two younger brothers and two younger sisters. Bernhard Schmidt apparently kept himself very much to himself even as a child, and had very little to do with his brothers and sisters, let alone other children. Like all the children on Naissar at that time Bernhard Schmidt attended a Swedish school. Nevertheless he still learnt German as a child, because his father had gone to a German-speaking school and continued to speak German to his children.

Bernhard Schmidt spent his childhood and formative years on Naissar or in Tallinn until the age of 22. No direct records remain from his school years. We can, however, be sure that he started experimenting as a schoolboy because at the age of 15 he suffered a terrible accident while experimenting with explosives. A tube packed with explosive blew up as he was holding it. His injuries were so severe that the lower part of his right arm had to be amputated. Bernhard Schmidt's determination overcame the consequences of this accident, because he carried out all his later professional activities and technical experimental developments with just his left hand. Moreover, his above-average experimental and practical skills become clear, because he had to make the majority of the tools that he needed to carry out his practical work, himself, and in accordance with his own special requirements. This situation, however, probably also served to encourage his solitary ways. He could not work with other people's tools, nor could other people work with his.

It is recorded that at the age of fifteen while still living on the island, Bernhard Schmidt built an electrical generator. He completed what was then a very complicated task using just his left hand.

At about the same age, Bernhard Schmidt was already interested in problems of optics. Using details in a textbook he built himself a camera, which he used for landscape photography. He also earned his first money with this home-made camera, by photographing people on the island and selling the pictures to them for small sums of money.

In order to earn a living after he had finished school, Schmidt moved from Naissar to Tallinn in 1895, and began work as a telegraphist for the local lifeboat service. The work did not seem to satisfy him, however, because between 1895 and 1901 he changed jobs several times in Tallinn. It is recorded that he worked as a photographer and retoucher, and finally in Tallinn's 'Volta' electromotor works.

During this time, however, he continued to follow his hobby of experimenting and it is clear that he was already concentrating more and more on working in the field of optics. In the *Astronomische Rundschau* for the year 1900, he was advised to buy himself a lens for his astronomical observations, because the five-inch lens he had made had such poor characteristics that it would not achieve much more than a good two-inch lens. This suggests that Bernhard Schmidt had already built himself a machine for grinding lenses for astronomical telescopes, whilst still living in Tallinn. He was not only concerned with building small astronomical telescopes at that time, because it can also be shown that he used them for his own observations. The *Astronomische Rundschau* mentioned him in 1901 as being one of the observers of the famous 'Nova Persei' that erupted in that year.

As a young man, Bernhard Schmidt was already interested in expanding his knowledge, and capabilities. This must be borne in mind considering his regular change of jobs in Tallinn as well as the fact that by the age of fifteen he was already using textbooks to build instruments. In 1900 he felt that he could no longer satisfy his needs in Tallinn and so he left Estonia for Sweden. He undoubtedly chose Sweden as the first place where he might further his knowledge, because of Sweden's influence on Tallinn and his own knowledge of the Swedish language that resulted from his attendance at the Swedish school on Naissar. The place of study that he chose was the Chalmes Institute in Gothenburg, where he registered as a student in 1901. His studies at this institute lasted, however, for a short time only, because he left Gothenburg again in October, 1901. It is thought that the course at the Chalmes Institute did not come up to his expectations and that his yearning for more practical knowledge for his experiments led him to search for new educational opportunities.

A technical college, the 'Technikum Mittweida', was founded in the small German town of Mittweida in 1867. At this establish-

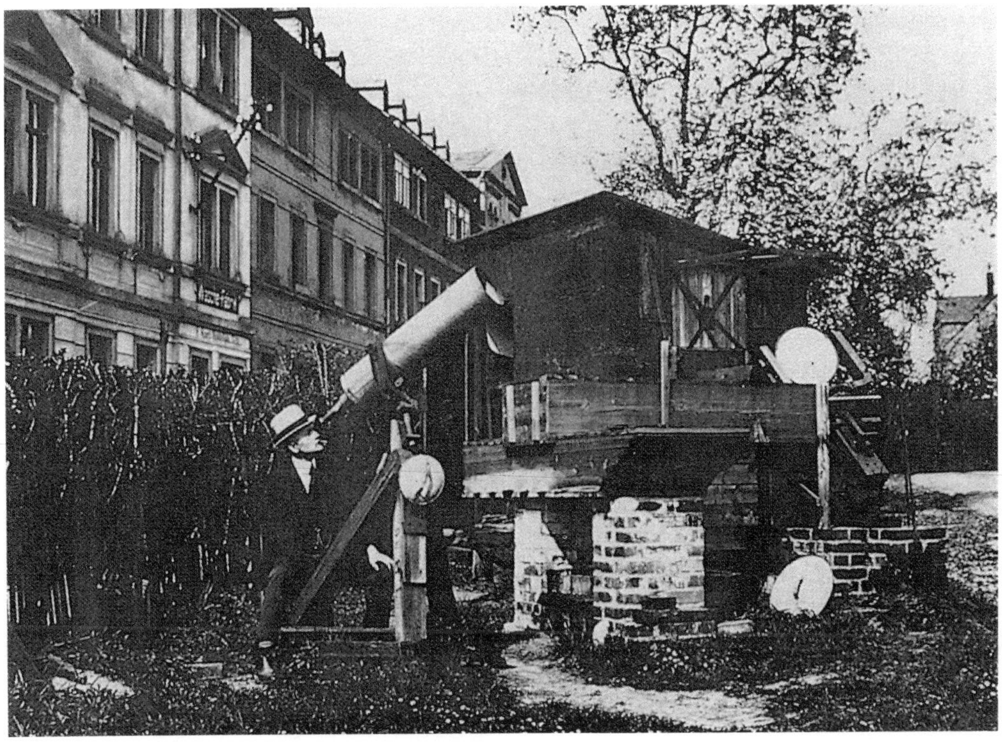

Fig. 5.2. Bernhard Schmidt not only produced astronomical optics but he also carried out astronomical observations immediately in front of his house in Mittweida, frequently in order to test the quality of his optical systems.

ment, laboratory technology had appeared as a subject as early as 1883. Electrical instruments and other devices were used here as teaching materials. As part of this development in Mittweida, the college's precision workshops were founded in 1901, and work began with the construction of electrical measuring instruments, motors and other laboratory equipment for the college and also for industry.

The college, now a school of engineering, was thus the first German educational establishment where laboratory work was part of the course and where the important relationship between theoretical and severely practical instruction was understood. This type of course must have undoubtedly attracted a man who was always intent on expanding his own knowledge and at the same time using it for the practical construction of equipment. It was logical, therefore, that Bernhard Schmidt should register for this course when he received informaton about the 'Technikum Mittweida' which was already internationally known. His knowledge of German certainly played a part too. When Bernhard Schmidt began his studies at Mittweida in October 1901, there were already many foreign students at the college, mostly from Russia.

The records of the engineering college tell us more about Schmidt's achievements as a student. In the first year he received 'good' and 'very good' marks for his achievements in mathematics, and in scientific and technical subjects. His results worsened considerably in the second year. In engineering design he only obtained 'satisfactory' and even, in the mechanical engineering course on steam engines and the construction of boilers, 'not satisfactory'. The records also show that he missed 75 double periods during the second year. This large number probably does not explain Bernhard

Fig. 5.3. An apparently earlier photograph showing one of Schmidt's many reflectors and the coelostat mirror for his horizontal telescope (which also appears on a more permanent mounting in the figure above*).*

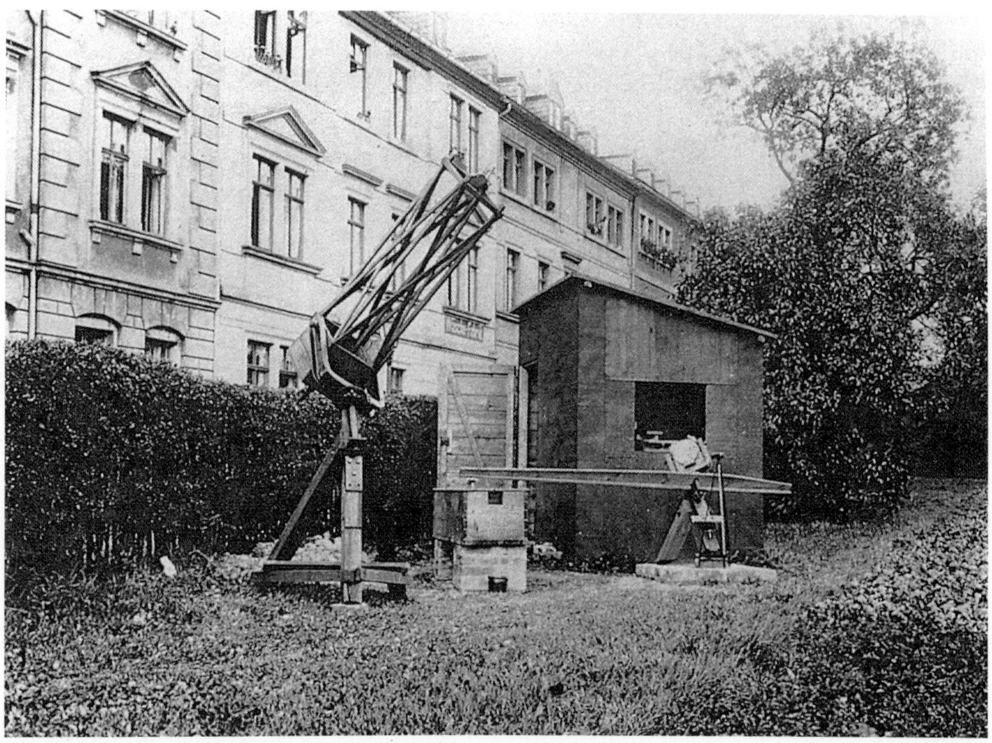

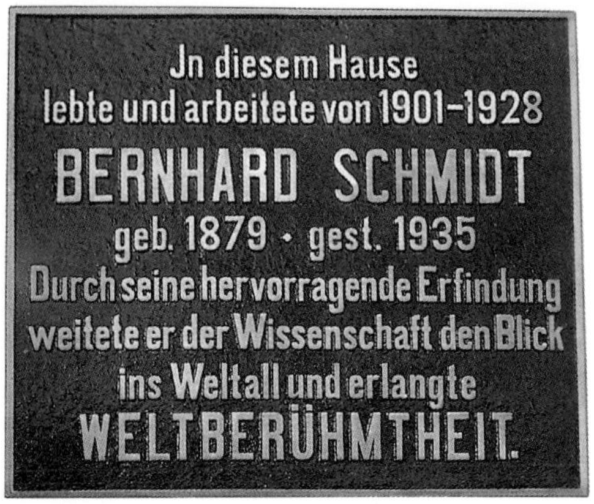

Fig. 5.4. A memorial plaque on Bernhard Schmidt's house shows when he lived in Mittweida and emphasizes the high esteem in which he is held in the town.

Schmidt's bad results, because he also missed a lot of lessons in the third and fourth years, when his results, with a few exceptions, were again 'good', and 'very good'. Considering Bernhard Schmidt's character, it is probable that he only worked hard in those subjects that would serve his own special interests. This is also supported by the fact that he wrote no dissertation and was therefore given no qualification on leaving. A stay at the internationally recognized, high-level and above all, practically orientated college, suited Bernhard Schmidt's interests in furthering his knowledge, but he was obviously not interested in acquiring an appropriate professional qualification.

There is another understandable explanation for the large number of lessons missed during the course. Bernhard Schmidt was already producing mirrors and lenses during his studies. This is implied, for example, by a letter from the director of the observatory that then existed in Altenburg, dated 12 August 1903, which can be found in the archive of the engineering college in Mittweida. It reads: 'The technician, Bernhard Schmidt, from I 4Z, spent his holidays with me, in order to make a parabolic mirror for me. The work is almost finished; a break now, with it being resumed in the autumn holidays would mean a great loss of time for us, because Mr Schmidt would have to become familiar with the work again in October, whereas he could finish the work now in just a few days. As Schmidt is a very talented man I believe I can assume that he would soon be able to catch up after a short absence from lectures, and I would therefore ask the directors of the college to excuse Schmidt from attending lessons for approximately eight days.' From this we can draw the conclusion that Bernhard Schmidt was indeed a talented student – a fact confirmed by the director of Altenburg Observatory, and also because the directors of the college granted him the days off, thus agreeing that he would catch up with his work.

In addition, his work for the Altenburg Observatory, which was only one example from many, makes it clear that Bernhard Schmidt already preferred practical work to other activities during his years as a student.

Finally, it shows that he also needed to work in order to ensure a financial income, because the course at the private college in Mittweida was expensive, and he can have expected very little financial support from his parents.

As a student, Bernhard Schmidt primarily made mirrors for amateur astronomers. The fact that this young, 22-year-old practical worker was already achieving outstanding results, never previously obtained, can be seen from a passage in a specialist magazine for astronomers in 1903: 'The virtuoso, Bernhard Schmidt, has carried out, as it were, a coup, in that he has made a mirror with the fantastic aperture ratio of 1 : 5. He deserves unstinting praise.' Mirrors with a large aperture ratio – and *f*/5 can certainly be considered large – have a relatively short focal length and require a highly accurate reflecting surface.

His workshops in Mittweida

After finishing his studies without writing a dissertation, what had been a sideline became his main source of income. Until then his accommodation and his 'workshop' had been identical, unless he was doing his grinding and polishing at the customer's premises, as at the Altenburg Observatory. When he was 25 years old, he applied to the Mittweida council for permission to set up an optical workshop. With this he had the basis for building up a business. True to his character he remained the only worker at the factory. He once stated his attitude very clearly as follows: 'Only one man alone is of any value, if several are together then there will be a quarrel. Among hundreds there will always be the riff-raff, and if there are thousands and more then they will start a new war.' This illustrates his character and his attitude towards people, which, although undoubtedly influenced by the time and society in general, was often somewhat negative.

By 1904 Bernhard Schmidt already had a good reputation as a maker of astronomical optics among many amateur astronomers, but as a self-employed worker he needed a larger clientele and, in order to fulfil their needs, a suitable workshop and machines. His first workshop was only a small shed; his second more roomy, situated in a house off a courtyard. The rooms of this second workshop still exist in Mittweida (Fig. 5.5), and in honour of Bernhard Schmidt, in 1956 the street in which they are located was named after him. The equipment in the workshop was very simple, virtually primitive. In order to grind mirrors that are rotationally symmetrical, the glass blank has to be rotated. In order to do this Bernhard Schmidt used a turntable, which he powered with his foot like a potter's wheel. He stuck the glass blank that had to be ground to the wheel with cobbler's wax. All the necessary tools were built to his specification by a local artisan in Mittweida.

Fig. 5.5. The equipment in Bernhard Schmidt's second workshop in Mittweida shows the simple methods he used to produce his lenses and mirrors.

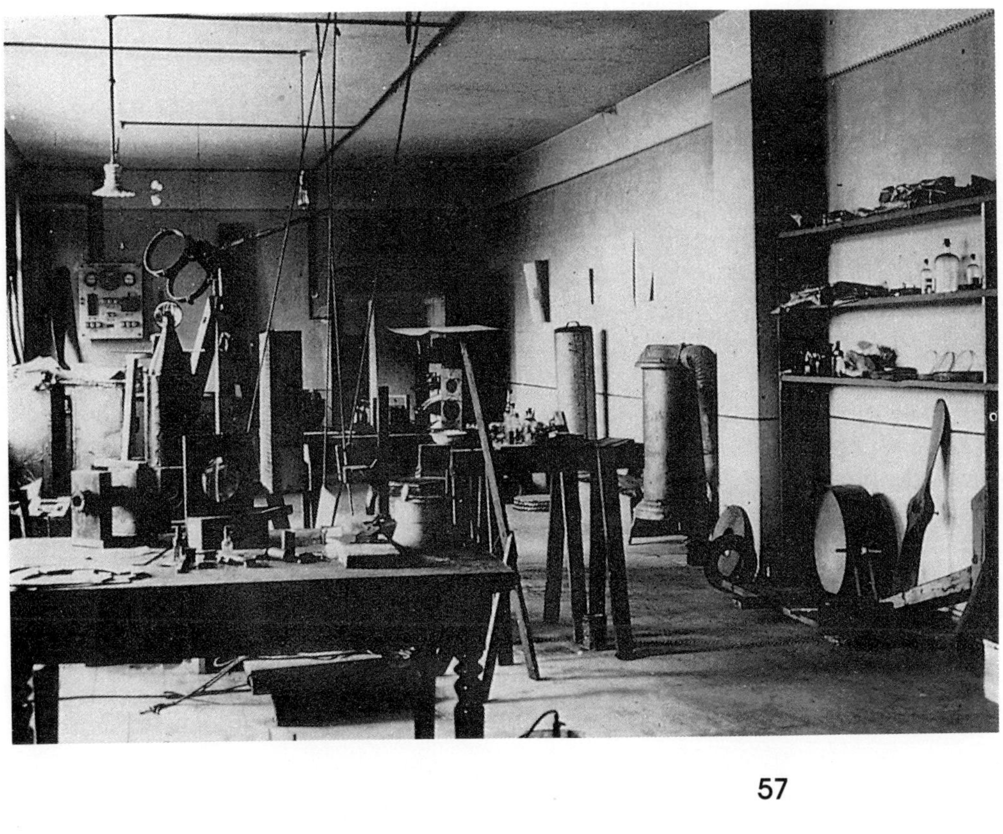

In order to find customers for his newly founded business, Bernhard Schmidt also contacted well-known observatories of the time. In May 1904 he wrote to the Astrophysical Observatory in Potsdam: 'Allow me to ask you whether you are interested in reflecting telescopes, I would very much like to supply a mirror at cost price.' He supported these offers with publicity material: 'I enclose a picture of the Moon taken with a 50-cm mirror (aperture ratio 1 : 10), and star trails with a 30-cm mirror (1 : 4) at full aperture. I guarantee the same visual definition as a refractor of the same aperture, but with larger dimensions.' In the original German, Schmidt made mistakes, but it has to be borne in mind that German was a foreign language to him.

He also offered his services to other well-known observatories. He wrote to the Director of the Hamburg–Bergedorf Observatory: 'Enclosed I allow myself to enclose several astronomical photographs that I have recently taken in Mittweida. . . . Also I allow myself to ask whether you have any astronomical optics that need improving, old lenses or suchlike. I am now able to obtain technical constants up to 0.1 on lenses, on demand.' The 'technical constant' was introduced by H. Lehmann in 1902. It is a measure for the average circle of confusion, which is produced by the overall zonal aberration of a lens or mirror, i.e. for minimum image size. In order to concentrate the incident radiation on a small area, one strives to achieve the smallest possible image size. The high quality that Bernhard Schmidt expected to attain with a technical constant of 0.1, becomes clear when one considers that, at that time, lenses with a technical constant of 0.5 were considered to be good. The object-glass of the largest refractor in the world, at the Yerkes Observatory has a technical constant of 0.16; the 76-cm object-glass of the famous Pulkova Refractor one of 0.18; and the Vienna 70-cm Refractor, made by the British firm of Grubb, one of 0.46. This offer also clearly illustrates that Bernhard Schmidt was sure of his ability. He was able, as the figures clearly suggest, to improve existing optics. The fact that Bernhard Schmidt was not successful immediately after founding his business is understandable, because respected optical firms already existed, such as the firm of Steinheil in Munich, and the firm of Grubb. It is therefore not surprising that the Director of the Potsdam Observatory stressed in a letter to Bernhard Schmidt in 1907: '. . . that any obligation to buy the mirror for the Institute is *not* implied by any test that may be carried out.' Bernhard Schmidt had sent a mirror to Potsdam to be tested as publicity, hoping that it would be purchased. The tone of cautious reserve in this letter indicates considerable doubt about the mirror before it was tested.

Even today the quality of astronomical optics is determined to a considerable extent by the test methods used, which provide information, as the optical surfaces are being made, of the accuracy that has been achieved. The process of making an astronomical mirror is a long series of sessions of polishing, testing, polishing again, and re-testing until the desired accuracy is attained, or until the testing methods being employed do not allow any improvement – because further improvements would require the detection of yet smaller deviations. The most accurate testing methods are therefore an essential part of producing astronomical optics. Bernhard Schmidt used the well-known Foucault knife-edge method to test the quality of his mirrors, the basic principle of which is still in use today. Another technique he used was *interference* testing.

Bernhard Schmidt also attempted to reproduce the actual conditions of use when carrying out his optical tests. He therefore made an effort to use point sources of light, like the stars themselves, when testing. To this end he used glass spheres, which reflected the sunlight, and which he mounted on long poles set up in Mittweida's town park. During the daytime, the reflected images of the Sun appeared as almost point sources.

When mirrors were completed, he took photographs of these artificial stars as a final quality control.

It is often maintained that Bernhard Schmidt achieved such high-quality mirrors through continuous testing, together with great skill as a craftsman. One magazine carried the following report about his work: 'Mr Schmidt's has recently tried to rescue the work of figuring, including that of mirrors, from hit-and-miss methods, and to produce a paraboloid by systematic calculation and techniques right from the start. He has succeeded in this attempt with almost mathematical accuracy.' This suggests that, just like the principle behind the course at Mittweida, he tried to combine both theory and practice in his work, and that he certainly did not produce his mirrors just by 'trial and error' and skill alone.

The high quality of Schmidt's work soon overcame the reservations about his claims. He proved the high quality of his work not only in making lenses and mirrors, but also in improving existing astronomical optics.

Karl Schwarzschild, H. C. Vogel's successor as Director of the Potsdam Observatory from 1909 to 1916 wrote, in connection with the necessary refiguring of one of the object glasses of the Potsdam twin refractor, to the Prussian Minister of Education on 7 June 1913: 'The undersigned most respectfully requests Your Excellency to hand over the correction work to Herr Schmidt. The undersigned is convinced that Herr Schmidt will be able to carry out the correction better and faster than the Steinheil company. Herr Schmidt is the greatest artist and as the matter concerns the completion of the largest German lens, the undersigned believes that in this connection all other considerations have to take second place.' We can certainly take 'other considerations' to mean reservations about giving a lens from the famous Steinheil company to a one-man business to be improved. This is, however, another indication of Bernhard Schmidt's ability. Karl Schwarzschild was certainly in a position to be able to judge the quality of the optics from Steinheil and Bernhard Schmidt's ability, because he had made extensive investigations of optical problems himself. Parabolic mirrors, which are the basis of all astronomical reflecting telescopes, produce perfect images only for rays that fall on to the paraboloid parallel to the optical axis. Only these rays converge at the focus free from aberration. Karl Schwarzschild had set himself the task of achieving a larger aberration-free field with a reflecting telescope, and had proposed a solution in which almost circular images were obtained over a field of nearly 3° in diameter with an aperture ratio of *f*/3.5. The fault with Schwarzschild's solution was that the images at the edge of the field had diameters of 16 seconds of arc, far larger than those close to the optical axis, where atmospheric seeing conditions caused them to have diameters of between one and three seconds of arc. Karl Schwarzschild's appreciation of Bernhard Schmidt's work must therefore be very highly rated.

On the 25th anniversary of the Astrophysical Observatory of Potsdam, which began operation on 1 June 1874, it was to be equipped with a large refractor, in particular because of its success in the field of stellar spectroscopy. The orders for this were placed in 1895 and the instrument was commissioned on 26 August 1899. It was a twin refractor with a lens of 80 cm diameter and 12 m focal length for photographic and spectroscopic work, and a lens of 50 cm diameter for visual observations. The optics were produced by the well-known firm of Steinheil in Munich, and the mechanical parts by the firm of Repsold in Hamburg, The first observations with this instrument were a great disappointment, especially where the optics were concerned. It proved possible for the optical aberrations of the 'photographic' objective to be reduced to a tolerable level after being retouched over a period of several years, but the anticipated limiting magnitude was never reached.

The 50-cm lens was given to Bernhard Schmidt for improvement. With some skilful retouching, he succeeded in turning it into one of the best lenses in existence. Thanks to its exceptional quality the famous astronomer, Ejnar Hertzsprung, who worked at the Potsdam Astrophysical Observatory laboratory from 1909 to 1919, and who made observations with the refractor after Bernhard Schmidt had worked on the lens, was able to make highly accurate measurements of the orbits of binary stars. Amongst other things, these observations served to determine stellar masses, one of the most important physical parameters of stars. Double-star observations with this instrument continued at Potsdam Observatory until 1967.

Bernhard Schmidt supplied astronomical institutes and amateur astronomers not only in Germany but also abroad. There are references to two astronomical mirrors from Bernhard Schmidt's workshop in the 1924 inventory of the Karls University in Prague. One is a parabolic mirror with a diameter of 60 cm and a focal length of 3 m, and a small convex mirror that belongs with it. The inventory records the price of this optical equipment as 13,016 crowns. The second mirror recorded has a diameter of 30 cm and a focal length of 4 m. The price was noted as 3238 crowns. This example shows that Bernhard Schmidt's abilities were known outside the boundaries of Germany.

For the first few years after starting his workshop in Mittweida, Bernhard Schmidt had a sufficient amount of orders. This was also the case throughout the First World War. At this period, he worked primarily for the Astrophysical Observatory in Potsdam. During the 1920s, business fell more and more into decline. In a letter dated 5 April 1925 to the director of the Hamburg–Bergedorf Observatory Schmidt complains that he was hardly receiving any orders and that he would prefer to sell the workshop in Mittweida, and perhaps start something new in a different town. The small number of orders for the production of astronomical optics was undoubtedly a result of inflation in Germany, and the general, world economic crisis, which was gradually looming on the horizon. Apart from this, the Carl Zeiss Works in Jena had developed to join the other Germany company of Steinheil in Munich, thus creating even greater competition for his one-man firm. Here we must consider the relationship that existed between the Carl Zeiss company in Jena and Bernhard Schmidt. The Jena company was perfectly prepared to employ Schmidt. This failed, however, once again because of Bernhard Schmidt's individuality. He demanded that his contract should state that he could decide on his own when he would work, and that he need only work when he felt like it. The company was not prepared to agree to these conditions and therefore did not employ Schmidt.

The most important well-known customers of Bernhard Schmidt's company in Mittweida were the Hamburg–Bergedorf Observatory and the Astrophysical Observatory at Potsdam. Relations with the Potsdam observatory ended with the death of Karl Schwarzschild in 1916. It is difficult to assess whether the reason for this was a lack of orders, or underestimation of Bernhard Schmidt's ability by Schwarzschild's successor as Director of the Astrophysical Observatory. Bernhard Schmidt was also in correspondence with Archenhold, the Director of the observatory at Berlin-Treptow. Letters written by Bernhard Schmidt show, however, that he was not satisfied with Archenhold's ability to keep promises. As a result no further co-operation took place.

His activities at Hamburg Observatory

The situation in which Bernhard Schmidt found himself in 1925 almost inevitably meant closer links with the Hamburg–Bergedorf Observatory. In 1927 he really did sell everything and move to Hamburg, where he became a voluntary worker at the Bergedorf Observatory. This description of

Bernhard Schmidt's job was used by Schorr, the Director of the Observatory himself. This indicates that Bernhard Schmidt had a lot of freedom at Hamburg that the Carl Zeiss company in Jena, for example, had not been prepared to allow him. Schorr, however, knew all about Schmidt's achievements. He could correctly estimate his ability and the amount of work he achieved, so he allowed him the status of 'voluntary colleague'. The results that Schmidt achieved in Hamburg proved how right Schorr had been.

In Bernhard Schmidt's obituary, Schorr described the task that he had given to his voluntary worker: 'My wish for our observatory to have a mirror with an aperture of 60 cm, an aperture ratio of at least *f*/2, and a large field of view led him to the construction of a coma-free mirror system.' The Director's wish represented a very high scientific goal, the solution to which was not achieved by Bernhard Schmidt in a single step.

In the 1930 Annual Report of the Hamburg–Bergedorf Observatory Schorr writes: 'Test exposures were carried out using the coelostat with a new fast coma-free mirror system, constructed by Bernhard Schmidt and consisting of a spherical mirror and a corrector plate located at the centre of curvature (Fig. 5.6). Bernhard Schmidt has published more detailed information about this system in *Central-Zeitung für Optik und Mechanik*, 52 (Vol. II). For a mirror already available here with an aperture of 44 cm and focal length of 62.5 cm, Schmidt has made a corrector plate with an aperture of 36 cm. Pictures taken with this system (*f*/1.75), e.g. Fig. 5.7, are encouraging: they show a completely coma-free field with a diameter of 15°.' Bernhard Schmidt had thus essentially achieved the considerable scientific and technical goals that had been set. The diameter of the mirror was only 44 cm instead of 60 cm, but what was critical was that a reflecting telescope was available that provided a 15 degree field of view. The coma-free field-diameter obtained by Bernhard Schmidt is approximately 30 times larger than any previously attained.

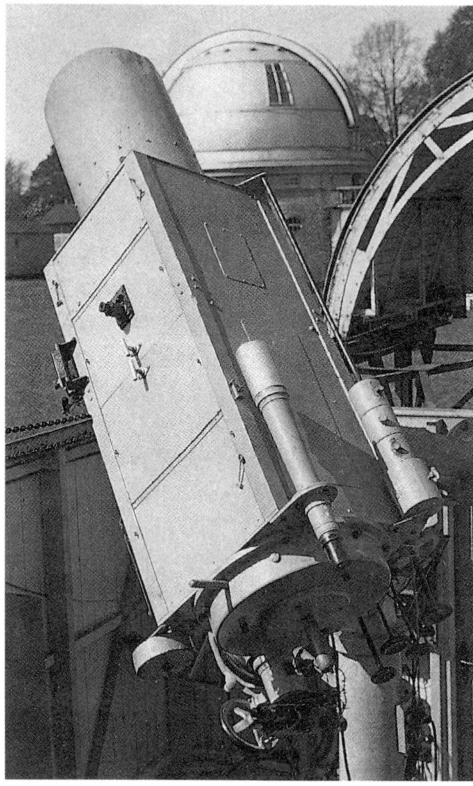

Fig. 5.6. One of the coma-free reflecting telescopes built by Bernhard Schmidt can still be found at the Hamburg Observatory.

Bernhard Schmidt published little about his work. He did, however, publish a short paper about his reflecting telescope in the *Central Zeitung für Optik und Mechanik*. This publication also appeared as *Mitteilungen der Hamburger Sternwarte in Bergedorf*, Nr. 36, and is reproduced here in the original (Fig. 5.8).†

In the 1930 Annual Report, Schorr speaks of 'encouraging exposures' with the new coma-free telescope. In *Vierteljahresschrift der Astronomischen Gesellschaft*, **70** (1935), Bengt Strömgren deals in great detail with the 'Schmidt reflecting telescope'. Here the new mirror system, its image quality, its residual aberrations, etc, are rigorously treated mathematically. Bengt Strömgren states, however, that excellent exposures had been taken with the Bernhard Schmidt mirror system and he concluded from his mathematical treatment of the new system '. . . that an exceptionally efficient, optical construction is now available'.

During the Second World War, the mirror of the first coma-free telescope produced by Bernhard Schmidt was unfortunately destroyed, and later had to be replaced by another. The corrector plate, however, is still the one made by its inventor.

In order that the principle of the coma-free, reflecting telescope could also be applied to larger mirrors with a greater image scale, Bernhard Schmidt produced a reflecting telescope with an aperture of 60 cm. This system had a focal length of 3 m. The Director of the Hamburg–Bergedorf Observatory wrote a report about this in the Annual Report for 1934. In order to compare the efficiency of the Schmidt reflecting telescope with the classical parabolic mirror, Schorr also had Schmidt produce a parabolic mirror with a diameter of 60 cm and focal length of 3 m. These two reflecting telescopes were set up at the Hamburg Observatory as a twin reflector on an English mounting.

Bernhard Schmidt also con-

†For a translation see Appendix.

Fig. 5.7. This picture of the North America Nebula taken by Bernhard Schmidt with his telescope in Hamburg emphasizes the size of the field of view and shows the excellent image quality right up to the edge of the field.

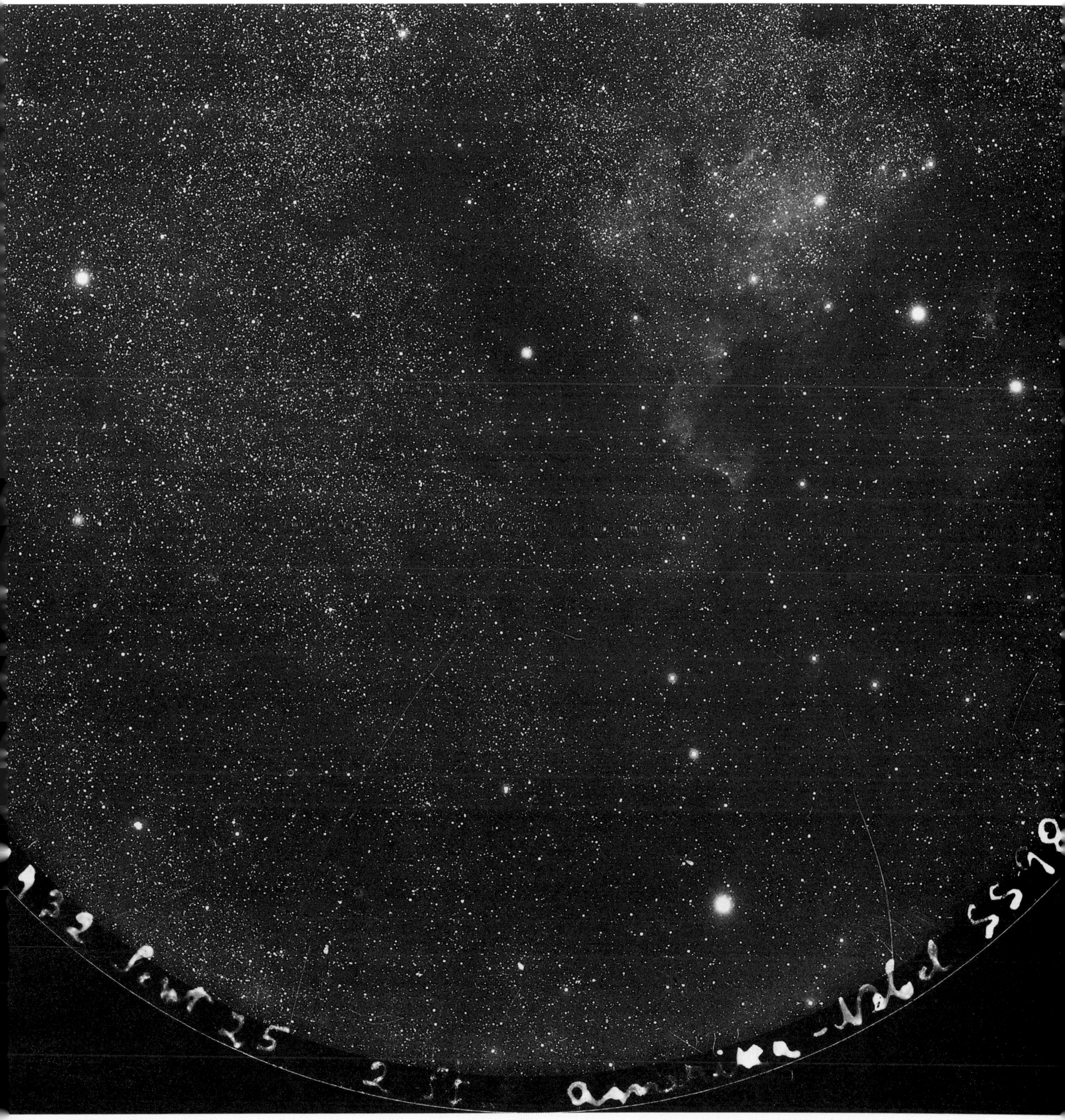

Mitteilungen der Hamburger Sternwarte in Bergedorf.

Band 7. Nr. 36.

Ein lichtstarkes komafreies Spiegelsystem*).

[Mit Tafel I und II]

Wenn man den Lichtverlust bei einem Spiegel und bei einem Linsensystem miteinander vergleicht, so ergibt sich, daß bei gleichem Öffnungsverhältnis der Spiegel einen geringeren Lichtverlust aufweist als das Linsensystem. Ein frisch versilberter Spiegel reflektiert mindestens 90% des auffallenden Lichtes, während ein Zweilinsensystem höchstens 80% und ein Dreilinsensystem höchstens 70% des einfallenden Lichtes durchläßt. Bei größeren Linsen wird durch die stärkere Absorption kurzwelliger Strahlen durch das Glas die Sache noch ungünstiger.

Bei großen Fernrohren wäre daher der Parabolspiegel im allgemeinen vorteilhafter als ein Linsensystem, leider wird aber bei großen Öffnungsverhältnissen das brauchbare Gesichtsfeld durch die Koma sehr beengt. Bei einem Öffnungsverhältnis 1 : 3 beträgt die Streuung durch Koma bei einem Gesichtsfelddurchmesser von nur 1 Grad bereits 37 Bogensekunden, außerdem kommt noch die Streuung durch Astigmatismus von 5 Bogensekunden hinzu. Die Koma wächst direkt proportional mit dem Gesichtsfelddurchmesser, der Astigmatismus quadratisch. Infolgedessen wird der Astigmatismus in der Nähe der Achse verschwindend klein und die Komaerscheinung tritt rein auf, während sie in größerem Abstand von der Achse durch den Astigmatismus modifiziert wird.

Immerhin ist der Parabolspiegel bei Öffnungsverhältnissen von 1 : 8 bis 1 : 10 dem gewöhnlichen Zweilinsenobjektiv in bezug auf Bildschärfe überlegen, hinzu kommt noch, daß eine Farbenabweichung beim Spiegel vollständig fehlt. Nachteilig ist, daß die Lichtverteilung in den Streuungsscheibchen der Spiegelbilder eine einseitige ist, was bei Ausmessungen derselben systematische radiale Verschiebungen hervorbringen kann.

Nun aber mag darauf hingewiesen werden, daß sogar der rein sphärische Spiegel mit den Öffnungsverhältnissen 1 : 8 bis 1 : 10 noch gut verwendbar ist. Würde man die Öffnungsblende direkt vor dem Spiegel anbringen, so würde gegenüber dem Parabolspiegel kein Vorteil entstehen, da ja der sphärische Spiegel genau dieselben Fehler hat; außerdem käme noch die sphärische Aberration hinzu, die über das ganze Gesichtsfeld die vorhandenen Streuungen vergrößert. Wird aber die Öffnungsblende im Krümmungsmittelpunkt angebracht, so hat der sphärische Spiegel, abgesehen von der Längsaberration, überhaupt keine Streuungen mehr, Koma und Astigmatismus sind null. Die Bildfläche liegt dabei auf einer Kugelfläche, die als Radius die Brennweite hat und konzentrisch zur Spiegelkrümmung ist, sodaß die Bildfläche mit der konvexen Seite dem Spiegel zugekehrt ist.

*) Central-Zeitung für Optik und Mechanik, 52. Jahrgang Heft 2.

Die Streuung des sphärischen Spiegels bei 1 : 8 bzw. 1 : 10 im paraxialen Bildpunkt beträgt 12.5 bzw. 6.4 Bogensekunden, die kleinstmöglicheStreuung nur ein Viertel davon, also 3.1 bzw. 1.6 Bogensekunden. Praktisch kann man sogar noch schärfere Bilder erhalten, wenn man zwischen diesen beiden Lagen einstellt. Unter normalen Verhältnissen sind diese Streuungen kleiner als die Eigenstreuung der photographischen Schicht. Daher ist auch bei Verwendung planer Platten die Bildgüte am Rande des Gesichtsfeldes besser als beim Parabolspiegel von entsprechendem Öffnungsverhältnis: die Sternbilder sind überall rund, mit symmetrischer Lichtverteilung.

Wird nun außerdem ein runder Planfilm durch Aufdrücken mit einem Ring auf eine der Bildfläche entsprechende Kugelfläche gekrümmt, was leicht ohne Faltung möglich ist, so sind auch noch die Zerstreuungsscheibchen im ganzen Gesichtsfeld von gleicher Größe. Man kann dasselbe auch mit einer scharfkantigen plankonvexen Sammellinse vor der planen photographischen Platte (plane Seite der Linse vor Platte) erreichen.

Wollte man aber das Öffnungsverhältnis stark vergrößern, so wird die sphärische Streuung sehr groß, da sie mit der dritten Potenz des Öffnungsverhältnisses wächst. Bei 1 : 3 oder gar 1 : 2 sind die Streuungen im paraxialen Bildpunkt 240 bzw. 800 Bogensekunden. Die kleinstmögliche Zerstreuungsscheibe hätte einen Durchmesser von 60 bzw. 200 Bogensekunden. Bei 1 m Brennweite wären dann die paraxialen Scheiben 1,2 mm bzw. 4 mm, oder die kleinstmöglichen 0,3 mm bzw. 1 mm groß. In diesem Fall wäre also der sphärische Spiegel nicht mehr verwendbar.

Nun will ich zeigen, wie man auch mit einem sphärischen Spiegel von großen Öffnungsverhältnissen noch vollständig scharfe Bilder erzielen kann.

Um aus einem sphärischen Spiegel einen parabolischen Spiegel herzustellen, muß man den Rand verflachen, ihm also einen größeren Krümmungsradius geben. Man kann aber auch auf den sphärischen Spiegel eine konzentrisch gekrümmte Glasschale (von überall gleicher Dicke) legen und deformiert deren eine Fläche. Nur müssen dann die Krümmungen umgekehrt sein, der Rand muß stärker gekrümmt werden als die Mitte. Auch muß der Deformationsbetrag etwa doppelt so groß sein. Denn jetzt erfolgt die Ablenkung durch Brechung; um gleiche Ablenkung bei Brechung wie bei Spiegelung zu erhalten, muß man ja etwa viermal so große Neigungen geben, da aber die Strahlen zweimal durch die Glasfläche gehen, so ist nur doppelt so große Deformation nötig.

Nun kann man auch noch diese Schale soweit optisch »durchbiegen«, daß die eine Fläche wieder plan wird, die andere Seite zeigt dann die reine Deformationskurve. Das heißt, man kann gleich von vornherein eine planparallele Platte deformieren. Optisch wird also im großen ganzen mit dieser Korrektionsplatte dieselbe Wirkung erzielt wie beim Parabolspiegel.

Eine solche passend geformte Deckplatte für einen sphärischen Spiegel hat auch den praktischen Vorteil, daß die Versilberung des Spiegels gut geschützt wird. Ein Nachteil ist der, daß infolge des zweimaligen Durchganges des Lichts durch die Glasplatte etwa 20% Lichtverlust auftritt.

Die Korrektionsplatte kann man auch an einer anderen Stelle in den Strahlengang einschalten. Befindet sie sich jenseits der Bildfläche, so geht das Licht nur einmal durch die Platte. Die Platte muß dann natürlich doppelt soviel Deformationswirkung haben wie im ersten Fall. Der Lichtverlust ist dann nur 10%.

Bringt man nun die Korrektionsplatte in den Krümmungsmittelpunkt des Spiegels, so ergeben sich wieder dieselben Verhältnisse wie vorher bei dem sphärischen Spiegel mit Öffnungsblende im Krümmungsmittelpunkt, nur mit dem Unterschied, daß jetzt auch die sphärische Aberration aufgehoben ist, und zwar über

Fig. 5.8. Bernhard Schmidt's sole publication about his famous optical system. (See Appendix for a translation.)

tributed his skills to the mechanical design of the twin reflector. Instead of the normal clockwork mechanism he devised and implemented a new type of drive.

In his work on the coma-free reflecting telescope, Bernhard Schmidt says nothing about the technique of making the corrector plate and so the process by which it was made was initially restricted to just the Hamburg–Bergedorf Observatory, which was obviously also in Bernhard Schmidt's interests. Nor is the method of producing the corrector plate mentioned in Bengt Strömgren's detailed study of 1935. The crucial information concerning the method by which the corrector plate was produced only reached the Carl Zeiss company in Jena after Bernhard Schmidt's death, when the firm's chief designer visited the Hamburg–Bergedorf Observatory in 1937. It is also known that one employee of the Hamburg Observatory gave some information about the production process during his stay in Pasadena in the USA. The fact that it was more than two decades after the invention of the coma-free reflecting telescope, before a significant number of Schmidt telescopes were built and used for astronomical research was not because astronomers were late in recognizing its significance. During the 1930s, particularly in Germany, but also in a number of other countries, problems that were not concerned with armaments

Mitteilungen der Hamburger Sternwarte Nr. 36 TAFEL I

AUFNAHMEN MIT DEM NEUEN SPIEGELSYSTEM SCHMIDT

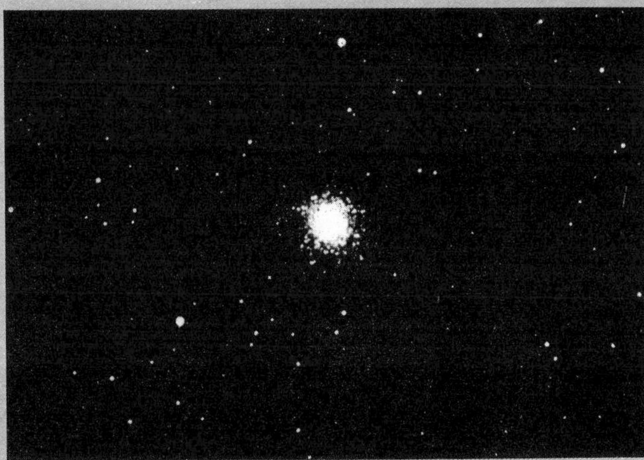

STERNHAUFEN IM HERKULES (M 13)
AUFNAHME IN DER OPTISCHFN ACHSE, BELICHTUNGSDAUER 40 MINUTEN
10FACHE VERGRÖSSERUNG

CYGNUS-GEGEND
10FACHE VERGRÖSSERUNG DER ANGEZEICHNETEN ECKE VON TAFEL II
ABSTAND VON DER OPTISCHEN ACHSE = 8°

das ganze Gesichtsfeld. Es ist also möglich, Öffnungsverhältnisse von 1 : 3 bis 1 : 2 zu benutzen und Freiheit von Koma, Astigmatismus und sphärischer Aberration zu erreichen.

Wenn die Neigung der einfallenden Strahlen sehr groß wird, so projiziert die Korrektionsplatte sich als Ellipse und die Deformation projiziert sich dann nicht auf die richtigen Stellen des Spiegels, sodaß die Korrektion sich ändert, und zwar tritt eine Überkorrektion in radialer Richtung ein.

Es kommen aber große Neigungen gar nicht in Frage, da ja sonst die photographische Platte größer würde als die freie Öffnung. Praktisch wird man kaum größere photographische Platten benutzen als von $^1/_4$ bis $^1/_3$ der Öffnung, und dann sind die Neigungsfehler noch verschwindend klein.

Etwas anderes ist es mit den chromatischen Fehlern der Korrektionsplatte. Um diese möglichst klein zu halten, wird man die Korrektionsplatte so gestalten, daß die mittlere Partie als schwache Sammellinse wirkt und die Randpartien eine Zerstreuungswirkung haben. Legt man die neutrale Zone in 0.866 des Durchmessers an, so ist die Chromasie ein Minimum. Ist die Umkehrstelle der Kurve bei 0.707, so ist die Randdicke der Platte gleich der Mittendicke. Die Dickenunterschiede zwischen den dicksten und dünnsten Stellen der Platte sind übrigens sehr klein, nur einige hundertstel Millimeter, sodaß eine störende Farbenwirkung nicht auftritt, jedenfalls ist die Wirkung viel kleiner als sonst das sekundäre Spektrum bei einem entsprechenden Objektiv.

Diese Chromasie ist identisch mit der sogenannten »chromatischen Differenz der sphärischen Aberration«.

Hat der Spiegel gleichen Durchmesser wie die Korrektionsplatte, so trifft der einfallende Strahlenzylinder bei den seitlichen Bildern exzentrisch auf den Spiegel und es bleibt ein Zweieck übrig, sodaß die Randpartien der Platte etwas weniger Licht erhalten. Will man das vermeiden, so muß der Spiegel einen größeren Durchmesser haben als die freie Öffnung, und zwar muß er um den doppelten Betrag des Plattendurchmessers größer sein. Bei einem Spiegel mit 50 cm freier Öffnung (Durchmesser der Korrektionsplatte) und 1 m Brennweite würde die photographische Platte bei einem Gesichtsfeld von 6 Grad einen Durchmesser von 10,5 cm haben, und demnach müßte der Spiegel einen Durchmesser von 71 cm haben.

Das hier beschriebene lichtstarke komafreie Spiegelsystem bietet nach den vorstehenden Ausführungen große Vorteile hinsichtlich Lichtstärke und fehlerfreier Abbildung. Voraussetzung hierfür ist aber eine technisch vollkommene Herstellung der Korrektionsplatte.

Bergedorf, 1931 Januar. BERNHARD SCHMIDT.

Zusatz. Herr *B. Schmidt* hat zu einem auf unserer Sternwarte vorhandenen sphärischen Spiegel von 44 cm Öffnung und 62.5 cm Brennweite eine Korrektionsplatte von 36 cm Öffnung hergestellt. Mit diesem System (1 : 1.75) hat Herr *Schmidt* eine Reihe von Versuchsaufnahmen bei horizontaler Aufstellung des behelfsmäßig hergestellten Kamerarohrs in Verbindung mit dem 60 cm-Zölostatenspiegel ausgeführt; von den auf gekrümmtem Film erhaltenen Aufnahmen sind auf Tafel I und II einige wiedergegeben. Man erkennt aus denselben, daß das Spiegelsystem trotz seines großen Öffnungsverhältnisses ein vollkommen komafreies Feld von 15° Durchmesser gibt, was für viele astronomische Untersuchungen einen großen Fortschritt bedeutet. Es ist beabsichtigt, für dieses Spiegelsystem ein besonders stabiles Kamerarohr herzustellen und dieses parallaktisch aufzustellen, um weitere Versuchsaufnahmen auszuführen.

Bergedorf, 1932 Januar. R. SCHORR.

received scant attention. This is why Bernhard Schmidt did not receive the recognition and honour that were his due during his lifetime.

The invention of the coma-free reflecting telescope is the best-known, and also the most significant, of Bernhard Schmidt's achievements. It would, however, give a one-sided view of his work if only this achievement were mentioned, because he also successfully examined many other astronomical and non-astronomical technical problems. Schorr wrote in Schmidt's obituary: 'Schmidt has also successfully investigated other technical and scientific fields such as aerodynamics, astrophotography, and production of large high-precision micrometer threads, and the construction of spectro-helioscopes.' Many of these technical problems that Schmidt solved were connected with the fact that he (according to Schorr) 'was also an enthusiastic astronomical observer'.

In order to observe, Bernhard Schmidt had built himself a horizontal telescope at Mittweida in 1909. This consisted of a parabolic mirror with a diameter of 40 cm and a focal length of 1 m, which was mounted on edge on a pillar in the open air. The axis of the mirror lay in a north–south direction. With a plane mirror, which could be turned around two perpendicular axes, Bernhard Schmidt directed the light from celestial objects onto the parabolic mirror. Accurate motors were required to move the plane mirror. The astronomers at the Hamburg–Bergedorf Observatory got to know of this horizontal telescope, and Bernhard Schmidt received an order to build a similar system for the Observatory. This was completed in 1920 and consisted of a plane mirror with a diameter of 61 cm, which directed light onto a parabolic mirror with a diameter of 55 cm and a focal length of 11 m, or else onto a second parabolic mirror with an aperture of 60 cm and a focal length of 30 cm. A feature of this horizontal telescope was that Bernhard Schmidt not only produced the optics, but also developed a special motor system. In the *Astronomischen Abhandlungen der Hamburger Sternwarte in Bergedorf*, **IV**, Nr. 1, 1928, J. Hartmann gives a detailed survey of the 'History and theory of astronomical instruments with a rotating plane mirror and with a fixed image direction'. In this he mentioned Bernhard Schmidt's observational equipment at Mittweida and also his system at Hamburg. Hartmann called such horizontal mirror systems 'Uranostats'.

Hartmann dealt in great detail with the theory of the motion of the plane mirror that is required to keep the light from a celestial object on the fixed, parabolic mirror as the sky rotates. This requires, amongst other things, a non-uniform rotational movement. 'A water-clock drive, as used by Bernhard Schmidt is very suitable for this because the speed can be altered very easily by altering the cross-section of the opening by which the water escapes.' Once again, Schmidt had succeeded in producing an exceptionally practical solution by using simple methods.

The Annual Report of the Hamburg–Bergedorf Observatory for the year 1928 states that Bernhard Schmidt himself carried out practical astronomical observations with his 'Uranostat' built for the observatory after his move to Hamburg, and that he took 'quite superb pictures of the planets, especially of Jupiter and Saturn, and of the Moon'. Later on, Bernhard Schmidt also supplied the Breslau Observatory with a similar horizontal telescope.

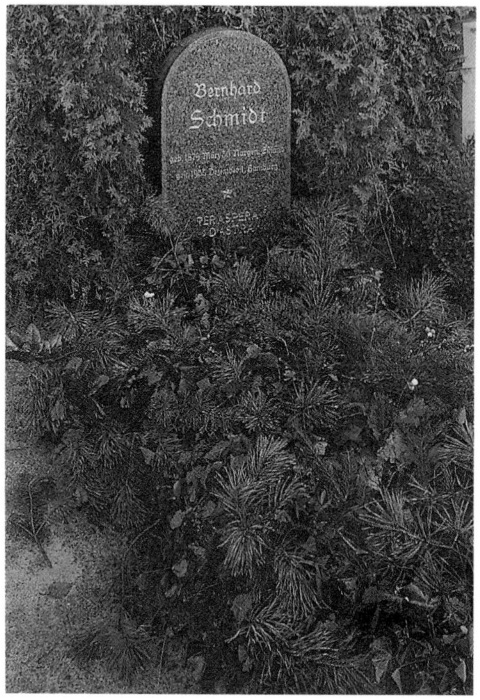

Fig. 5.9. Bernhard Schmidt's grave is close to the building of the Hamburg Observatory.

In Mittweida, Schmidt also made observations of the Sun and took photographs of sunspots with a mirror having a focal length of 9.4 m. These caused general surprise amongst astronomers at the time because they were exceptionally sharp.

In Mittweida, and later in Hamburg, Bernhard Schmidt worked intensively on problems that had nothing at all to do with astronomy. In a letter dated 23 September 1926, he wrote the following from Greifswald to Schorr, the Director of the Hamburg Observatory: 'I have already made such progress with my sail experiments that I can sail forwards directly against the wind by wind power alone, admittedly not very quickly, but all the same I can reach the target faster than a sailing boat that has to continuously tack against the wind. Running before the wind, I can travel faster than with normal sails.' Bernhard Schmidt was undoubtedly prompted in this activity by his childhood years on the island of Naissar, where he had often experienced that the relatively long stretch of water from Naissar to Tallinn on the mainland had to be crossed with a lot of strenuous rowing. Once again, when building this 'head-wind boat', Bernhard Schmidt developed the theory and then immediately put it to practical use. His theoretical calculations showed that a 'head-wind boat' could reach a speed against the wind that amounts to 25% of the wind speed.

A letter from Bernhard Schmidt from Naissar to the director of the Hamburg Observatory in 1930 makes it clear that he continued intensive work on the 'head-wind boat' after he had become a voluntary worker at the Observatory, and that for this purpose he had returned to Naissar for a long period. He wrote: 'In the autumn I also made test journeys here of 20 km, between Nargan and Reval with my 'sail-mill', I can reach much better speeds than with normal sails.' His success in both theory and practice led Schmidt to apply for a patent for his 'head-wind boat' in 1927. A positive report is supposed to have been written by the well-known German aerodynamics expert, Ludwig Prandtl, from Göttingen. Despite this, Bernhard Schmidt's 'head-wind boat' was not recognized as a patent. A decision of this sort certainly contributed towards the fact that the loner became more and more withdrawn.

Bernhard Schmidt died on 1 December 1935, in Hamburg –Bergedorf, bitter and depressed. He was buried very close to the

Observatory. Although Bernhard Schmidt had always worked alone, had had little contact with other people and was known as a loner and an eccentric, he could also be cheerful in company. Schorr, who had known him for a long time and very well, wrote the following in his obituary: 'Personally Schmidt, who was unmarried, was an eccentric, and a somewhat reserved man, but who, on the other hand, in pleasant company, could be cheerful himself. To the employees at our observatory he was a well-liked, and always helpful colleague.'

6 Astronomical objects in words and pictures

This section shows photographs of 33 astronomical objects, taken with Schmidt telescopes. As 'attractive' objects were wanted the largest number of exposures show various forms of interstellar material, followed by photographs of galaxies.

Detailed information about these pictures can be found in Table 6.1 given here. Apart from the name of the objects, the serial numbers in the *New General Catalogue* (NGC) and in the *Messier Catalogue* are also given, if they exist. The coordinates in right ascension and declination for Epoch 2000.0, are also given, together with the date of exposure and the exposure time. Information is then provided concerning the colour in which the respective exposure was made (U = Ultraviolet, B = Blue, R = Red). In order to give some impression of the size of the areas shown, the length of the shorter side of each picture is given as an angular measurement in the column headed 'field size'. The arrow in the next column indicates the direction of North. The Schmidt telescope used for each exposure is also indicated (KSO = Schmidt telescope of the Karl Schwarzschild Observatory; UK = United Kingdom Schmidt telescope in Australia; Jena = University Observatory in Jena; Hbg = the original Schmidt Telescope at the Hamburg Observatory).

Table 6.1.

Object	Right Ascension (2000)	Declination (2000)	Date	Exposure time	Colour band	Field size(°)	Orientation	Telescope
	h m							
Star field	00 09.0*	61° 00′*	1964 Sep. 14/15	75 Min.	B	2.1	←	Jena
Andromeda Galaxy NGC 224 = M31	00 42.8	41 11	1984 Aug. 27/28	25	B	1.8;0.7	↑ ↑	KSO
Triangulum Galaxy NGC 598 = M33	01 33.9	30 40	1985 Sep. 11/12 1985 Sep. 14/15 1963 Sep. 17/18	23 75	B U R	0.7 0.5 0.5	↑ ← ←	KSO KSO KSO
Open clusters h and χ Persei NGC 869, NGC 884	02 19.1 02 22.5	57 09 57 07	1962 Jan. 05/05	60	V	1.0	←	KSO
Pleiades M45	03 46.9	24 07	1985 Oct. 21/22	34	B	1.0;1.2;1.2	↑ ↑ ←	KSO
California Nebula NGC 1499	04 03.3	36 25	1975 Feb. 08/09	55	R	1.6	←	KSO
Magellanic Clouds	05 20.0 00 50.0	−69 00 −73 30	–				← ←	UK
Crab Nebula NGC 1952 = M1	05 34.5	22 01	1964 Feb. 09/10	63	R	0.2	↑	KSO
Orion Nebula NGC 1976 = M42	05 35.5	−05 25	1975 Feb. 09/10	55	R	1.6	↑	KSO
Horsehead Nebula NGC 2023	05 41.8	−02 13	1972 Jan. 20/21	60	R	1.3	↑	KSO
Open clusters NGC 2158 NGC 2168 = M35	06 07.3 06 08.8	24 06 24 21	1964 Jan. 20/21	15	B	0.5	←	KSO
Rosette Nebula NGC 2238–2239	06 30.3	04 59	1972 Jan. 15/16	80	R	1.6	←	KSO
Hubble's Variable Nebula NGC 2261	06 39.2	08 44	1972 Jan. 18/19	50	R	0.8	↑	KSO

Table 6.1. (*cont.*)

Object	Right Ascension (2000)	Declination (2000)	Date	Exposure time	Colour band	Field size(°)	Orientation	Telescope
Cone Nebula	06 41.0	09 54	1972 Jan. 18/19	50	R	1.8;0.3	↑ ↑	KSO
Galaxy M 81 NGC 3031	09 55.8	69 04	1976 Feb. 28/29	40	B	0.6	↑	KSO
Eta-Carina complex NGC 3372	10 53.8	−59 25	–				↑	UK
Coma Cluster of galaxies	12 56.0*	28 33*	1972 Feb. 21/22	30	B	0.8	←	KSO
Virgo Cluster of galaxies and M87 = NGC 4486	12 30.9	12 24	1970 April 12/03	18	B	0.8;0.5	←↑	KSO
Objective-prism plate	12 30.9	12 24	1980 Mar. 10/11	8		0.8	←	KSO
Centaurus-A NGC 5128	13 25.3	−42 45					↑	UK
Whirlpool Galaxy NGC 5194 = M51	13 29.9	47 +11	1961 Feb. 17/18	60	U	0.4	↑	KSO
Galaxy M101 NGC 5457	14 03.3	54 22	1962 Feb. 09/10	105	R	0.5	↑	KSO
Globular cluster NGC 6341 = M 92	17 17.2	43 08	1966 Sep. 20/21	30	B	0.2	↑ ↑	KSO
Lagoon Nebula NGC 6523 = M8	18 03.1	−24 23	1972 Jun. 08/09	50	R	1.8	←	KSO
Omega Nebula NGC 6618 = M17	18 20.7	−16 12	1969 Jun. 09/10	55	R	0.8	←	KSO
Dumbbell Nebula	19 59.6	22 43	1962 Oct. 03/04	25	B	0.3	↑	KSO
Area around γ Cygni	20 27.7	40 18	1963 Aug. 20/21	30	R	1.7	←	KSO
The Cygnus Loop	20 45.6	30 43	1986 Jly. 02/03	33	U	3.1	↑	KSO
NGC 6960	20 45.6	30 43						
NGC 6992	20 56.3	31 42						
NGC 6995	20 57.0	31 43						
Pelican Nebula IC 5067, IC 5070	20 48.6	44 22	1969 Jun. 08/09	90	R	1.1	↑	KSO
North America Nebula NGC 7000	21 01.9	44 12	–			5.9	↑	Hbg
North America Nebula NGC 7000	21 01.9	44 12	1969 Jun. 08/09	90	R	1.8	←	KSO
Coccoon Nebula IC 5146	21 53.3	47 16	1978 Oct. 08/09	40	R	0.5	←	KSO
Star trails around the Celestial Pole		90 00*	1968 May 02/03			1.8		KSO
Comet Bennett	22 27.4	24 56	1970 Apr. 02/03	24	U	1.4		KSO
	23 01.4	43 20	1970 Apr. 11/12	40	U	1.6		KSO
	23 01.4	43 20	1970 Apr. 11/12	35	R	0.7		KSO
	23 05.9	44 50	1970 Apr. 12/13	30	U	0.7		KSO

*The coordinates given are for the centre of the field

A general star-field

In this photograph of a star-field there are no particular objects, such as bright interstellar regions, large galaxies or such like. The only noticeable differences are the various diameters of the stellar images, and the increase in the number of stars as the images get smaller.

The size of the stellar images is a measure of the brightness of the stars. All stars are so far away that we can only observe them as points of light with no measurable extent. Even if the Sun were to be moved to the distance of the nearest star, Proxima Centauri, and had a diameter ten times as great, its angular diameter would still be below one-tenth of a second of arc. This lies below the resolving power of telescopes, but, in particular, is well below the 'resolving power' of the Earth's atmosphere. Stars nevertheless appear of different sizes on photographs because with bright stars, a large amount of light falls onto the photographic layer and is then spread out by scattering in the emulsion.

In ancient times, the stars were grouped into magnitudes according to their brightness. The brightest were called first-magnitude stars, the next brightest: second-magnitude stars, etc, and those stars that could just be discerned with the naked-eye, sixth-magnitude stars. When the telescope become available, it was necessary to extend this scale beyond the sixth magnitude. In the last century, widely-used, extended scales were introduced by F. W. Argelander, W. Struve and W. Herschel. There were, however, considerable differences between their scales. A star of thirteenth magnitude on Argelander's scale was magnitude 11.5 on Struve's and 18 on Herschel's.

In order to be able to compare observations by different authors with one another, a uniform scale is required. An examination of the available scales showed that, on the Argelander scale, the intensity ratio between two stars one magnitude apart was indeed subject to distinct variations but, on average, amounted to approximately 2.5. Because of the particular properties of the human eye there is a logarithmic relationship between the intensity of the stimulus and the sensitivity as expressed by the magnitude scale. The value quoted also enables historical data to be handled simply from a numerical point of view.

As already mentioned, under reasonable atmospheric conditions it is possible to see sixth-magnitude stars with the naked eye. The use of modern observational techniques has enabled fainter and fainter stars to be detected. With the largest Schmidt telescopes stars of about magnitude 22 have been recorded. The effective radiation flux received from such stars is less than one two-millionth of that from a sixth-magnitude star, or about one three-millionth of that from the brightest star, Sirius, which has a magnitude of −1.45. This is another indication of the light-grasp of large Schmidt telescopes.

As can be seen from the photograph, the number of stars increases markedly with decreasing brightness.

In the middle of the last century, Argelander recorded all the stars in the northern hemisphere down to magnitude 9. According to these charts and catalogues, which are known as the 'Bonner Durchmusterung', in the area between the North Celestial Pole and the celestial equator there are nine stars above magnitude 1, 30 stars between the magnitudes 1 and 2 – therefore giving a total of 39 – and a total of 2883 down to magnitude 6. Argelander recorded a total of 203,649 stars above magnitude 9 in the northern sky. If the increase from one magnitude to the next is considered it is found to be approximately four-fold. This can also be predicted theoretically, and, making certain assumptions, a value of 3.982 is obtained for the increase in the number of stars per magnitude interval.

If we assume that this rate of increase is valid down to magnitude 18 (for example), this means that there are between 40 and 50 thousand million stars above this magnitude in the northern sky. It is, however, by no means possible to observe all these, because, with increasing distance interstellar extinction becomes greater, causing the number of visible stars to increase more slowly, or even remain constant with the distance. Near the galactic equator, where the density of stars per area of sky is greatest, there are, on average, 10,000 stars per square degree down to magnitude 18. If the field of a Schmidt telescope is about 10 square degrees, this means that there are about 100,000 stars above that magnitude on each plate. This indicates the wealth of information that can be gained by telescopes of this type.

Fig. 6.1. A field of stars in the constellation Cassiopeia.

The Andromeda Galaxy

In the constellation of Andromeda a small, diffuse, luminous spot can be seen with the naked eye close to the star ν Andromedae. This was the 31st object recorded by Messier in his catalogue. One of the first descriptions of it comes from Marius, in 1612 when he compared this object's glow with the light of a candle shining through a horn lantern. There are also indications to suggest that in the Middle Ages the Arabs had already observed this diffuse spot near the star ν Andromedae.

'Rarely have views about the true nature of a celestial body been so changed as they have in the case of this nebula as a result of photography' is a very justifiable statement made in the 5th edition of *Popular Astronomy* by S. Newcomb and R. Engelmann in 1914. The 1914 *Popular Astronomy* also contains the following: 'Pictures of the nebula by various observers, in particular Roberts, the Lick astronomers, and Ritchey, agree in showing that the decrease in light towards the edge of the nebula is by no means uniform, and that the nucleus is surrounded by strong elliptical rings, which at first sight appear very similar to the Saturn system. On closer examination, however, one can see that they are not separate concentric rings but that the nebula is a huge spiral nebula, similar to the one in Canes Venatici, with the difference that here one sees the system almost edge-on.' This 1914 description contains the essential facts about the Andromeda Nebula that have been confirmed subsequently.

A decisive new observational result was obtained in 1923 by Edwin Hubble when he was able to prove that the distance of the Andromeda Nebula is so great that it does not belong to our Milky Way system, but is an independent extragalactic stellar system. Determining the distance of the Andromeda Galaxy became possible when modern observational technology allowed individual stars to be detected within it and their absolute magnitudes could be determined, for example, by virtue of their variability. By comparing the absolute brightness with the observed apparent brightness, it is possible to determine the distance. As a result, the Andromeda Nebula has a distance of 7×10^5 pc. In the universe as a whole this is a very small distance because there are also stellar systems that we know to be nearly 2×10^{10} pc away.

Just as the Sun, the nearest star, can provide us with a lot of information about stars, the Andromeda Galaxy has been studied intensively in order to gain general information about galaxies.

Investigation of the Andromeda Nebula is made much more difficult because its flattened disc-like shape is seen at an inclination of 77°, causing the various structures to appear distorted. However, we can clearly see a bright central region, with a diameter of 5000 pc, surrounded by fainter regions. On very short-exposure photographs, a star-like nucleus can be seen in the centre of the bright central region, and this has a diameter of only 10 pc. The infrared luminosity of the nucleus of the Andromeda Galaxy is nearly as great as that of the centre of our Milky Way system. From this, we may conclude that the stellar composition of both nuclei is almost identical. The flux of radio emission from the nucleus of our Galaxy is almost 20 times greater than that from the nucleus of M31, however. It is thought that the small nucleus of the Andromeda Galaxy contains ten million solar masses. The density of stars within it is therefore 10,000 times greater than in the vicinity of the Sun.

Optical and radio observations of the Andromeda Galaxy show that it has a total extent of somewhat more than 25,000 pc. Photographic observations of the region around the bright central core have revealed Delta-Cephei stars, about 1000 regions of ionized hydrogen, and approximately 200 stellar associations consisting of young, hot stars. These objects are typical of spiral arms. An exact examination of the distribution of the spiral arm indicators in the Andromeda Galaxy makes it clear that it is a spiral system and, like our Galaxy, can be classified among the Sb-spirals.

Interesting results have been gained from radio observations of the Andromeda Galaxy. One cause of radio-frequency radiation from galaxies may be electrons, which spiral around interstellar magnetic-field lines, giving rise to what is known as synchrotron radiation. Another cause could be free-moving electrons in a plasma, whose motion is determined by the local energetic equilibrium. This type of radiation is therefore known as thermal emission or bremsstrahlung. The two types of radiation may be distinguished by their spectral composition. In the case of synchrotron radiation the intensity decreases with decreasing wavelength. Bremsstrahlung is almost independent of the wavelength and wavelengths of more than ten centimetres predominate with this type of radiation. The radio-frequency emission observed in the Andromeda Galaxy is mostly of a non-thermal nature, i.e. synchrotron radiation. This means that the Andromeda Galaxy must have a magnetic field and that high-velocity cosmic-ray electrons must be present. Sources of these electrons could be supernova remnants, flare stars, and X-ray sources. Only 25% of the radio-frequency radiation can be attributed to thermal sources. These include the regions of ionized hydrogen.

Using Doppler-shift measurements it has been shown that the Andromeda Galaxy rotates. As a result various rotation zones may be distinguished. The tiny, star-like nucleus rotates very rapidly. A velocity of nearly 90 km/s has been measured at its outer edge. As

Fig. 6.2. The Andromeda Galaxy (M31).

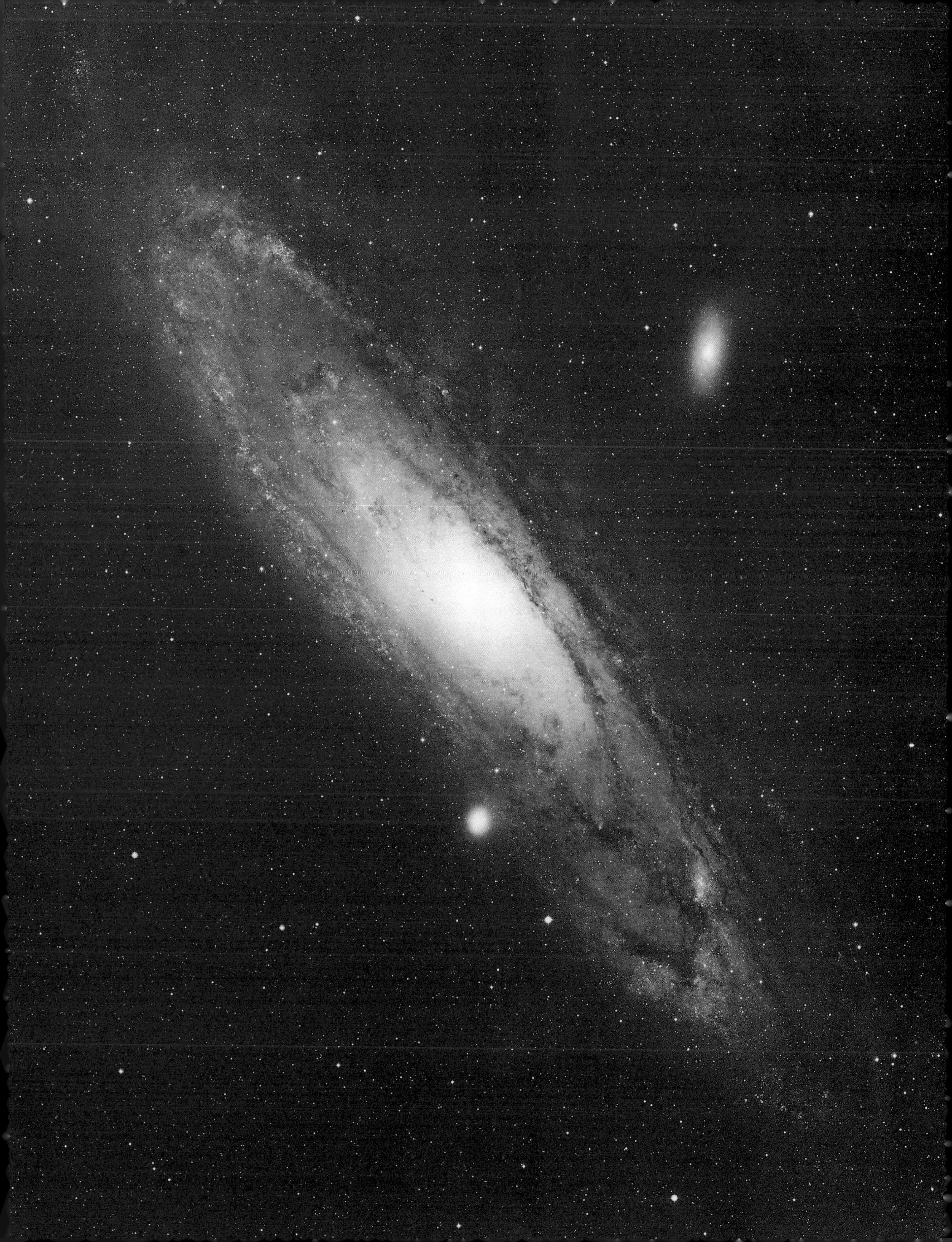

Table 6.2. *Members of the Local Group (a partial list)*

Galaxy	Type	Diameter in pc	Distance in pc	Mass in solar masses	Radial velocity in km/s
M31	Sb	30,000	675,000	3.1×10^{11}	−300
Galaxy	Sb	30,000	–	1.4×10^{11}	–
LMC	Irr	7,100	50,000	1×10^{11}	+270
M33	Sc	6,100	740,000	13×10^{9}	−190
NGC 147	E 5	1,000	700,000	10×10^{9}	−250
NGC 185	E 3	1,000	700,000	10×10^{9}	−300
NGC 205	E 5	2,000	675,000	8×10^{9}	−240
M32	E 2	1,000	675,000	3×10^{9}	−210
SMC	Irr	3,100	52,000	2×10^{9}	+170
NGC 6822	Irr	2,000	460,000	3.2×10^{8}	−40
IC 1613	Irr	1,000	740,000	3×10^{8}	−240
Fornax	E 3	2,000	185,000	2×10^{7}	+40
Leo I	E 3	1,000	230,000	4×10^{6}	
Sculptor	E 3	1,000	89,000	3×10^{6}	
Leo II	EO	1,000	230,000	1×10^{6}	
Draco	E 3		62,000	1×10^{5}	
UMi	E 5		62,000	1×10^{5}	

the distance from the centre increases, the rotational velocity first decreases, and then increases again, to a new maximum of 100 km/s at a distance of 700 pc from the centre. It then decreases again, but subsequently rises to reach a velocity of 300 km/s – the greatest velocity recorded in the Andromeda Galaxy – at a distance of 15,000 pc from the centre. From here, the velocity decreases slowly towards the edge of the galaxy. Conclusions can be drawn from the determination of the rotational velocity as to the mass of the stellar system. This leads to a mass of 310 thousand million solar masses, about double that of our Galaxy. The mass of interstellar matter in the Andromeda Galaxy may be determined from the radio observations already mentioned. It amounts to only about 1%, similar to the amount in the Milky Way system, which amounts to about 2%.

About 200 globular clusters have been found in a spherical halo surrounding the flattened disc of the Andromeda Galaxy. These clusters are the oldest members of the Andromeda Galaxy, which again corresponds with the structure of our Milky Way system. The Andromeda Galaxy and our Galaxy share so many similarities that they can definitely be described as twin systems.

The Andromeda Galaxy is, of course, also taking part in the systematic expansion of the cosmos. However, because the distance between the Andromeda Galaxy and the Milky Way system is very small, the expected velocity of recession of the Andromeda Galaxy is also very small and should amount to only about 50 km/s. Apart from this systematic motion, all galaxies have an individual motion and, in the case of nearby galaxies this may be greater than the systematic recession velocity. This is also true for the Andromeda Galaxy. Its radial velocity towards the Milky Way is considerably greater than the velocity of recession, so the Andromeda Galaxy is approaching us at a velocity of 300 km/s.

Two other galaxies can clearly be recognized on the photograph of the Andromeda Galaxy. These are the galaxies M32, immediately next to the edge of the Andromeda Galaxy, and NGC 205 somewhat farther away. Both of these show no signs of spiral structure and are classified as elliptical galaxies. When their masses are compared with that of the Andromeda Galaxy they are found to be dwarfs. M32 has a mass of only 3×10^{9} solar masses, and NGC 205 8×10^{9} solar masses. Their diameters, 1000 and 2000 pc, respectively, are also significantly below those of the Andromeda Galaxy and our Galaxy. They are therefore classed as dwarf elliptical galaxies.

Both of these galaxies are really very close to the Andromeda Galaxy in space and they can therefore be regarded as close companions of the large Andromeda Galaxy.

A detailed examination of the distribution of galaxies shows that they normally appear in clusters and groups. Investigation of the vicinity of our Milky Way system has shown that the same applies here, and that about 30 galaxies form a unit, which is known as the Local Group (Table 6.2).

Where there are gaps in the table, no reliable information has yet been collected. Positive radial velocities indicate that the system is moving away from the Milky Way system, and negative velocities mean they are approaching the Milky Way. This representative selection emphasizes the dominance of the Milky Way and of the Andromeda Galaxy, M31, and shows that the great majority of the members of the Local Group are dwarf systems. This also indicates the importance of studying the Local Group because dwarf systems of only 100,000 solar masses like the Draco system, for instance, cannot be detected in distant clusters of galaxies.

Fig. 6.3. The southern portion of the Andromeda Galaxy.

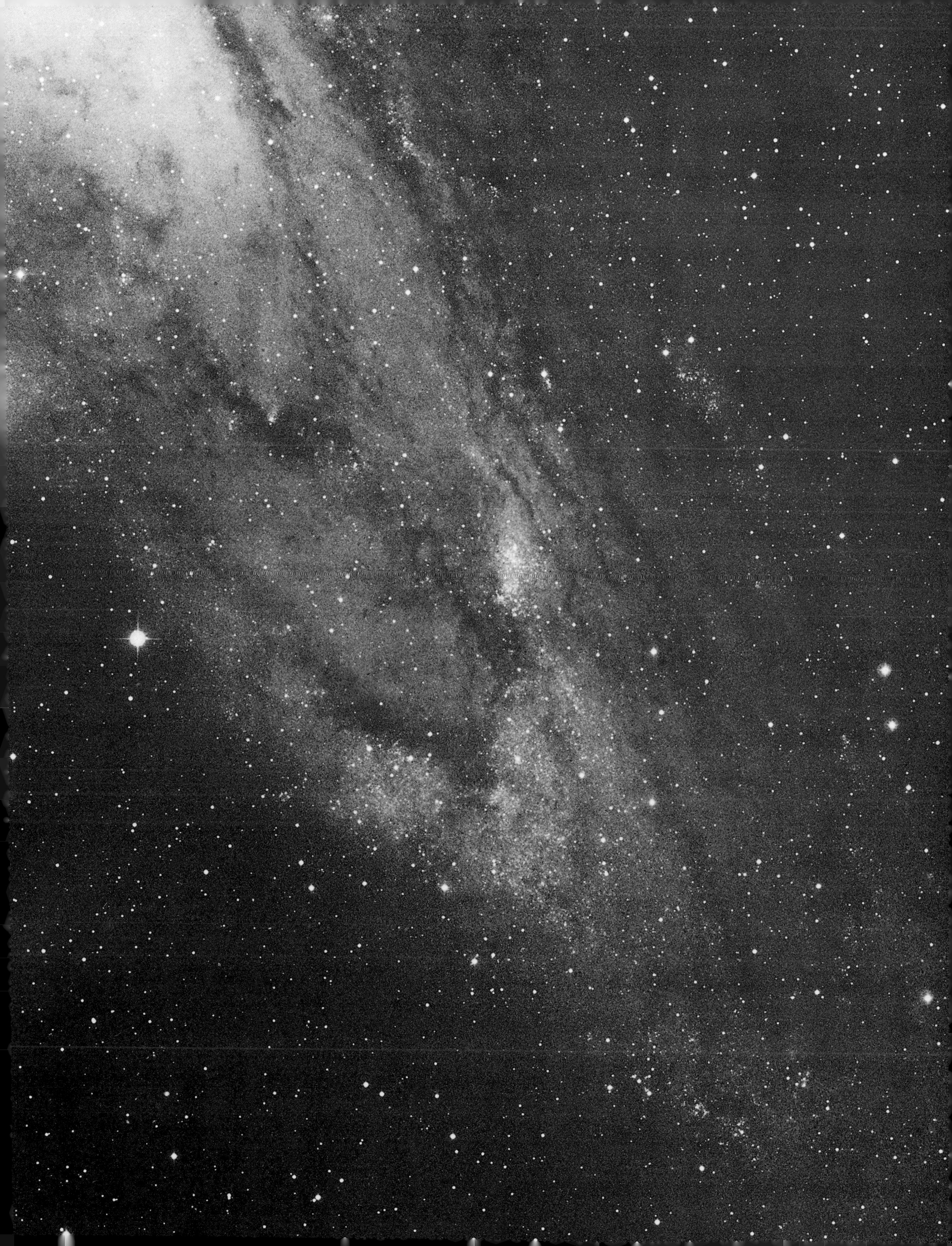

M33

The galaxy listed as number 33 in the *Catalogue of Nebulae and Star Clusters* compiled by Charles Messier, who lived from 1730 to 1817 and published the first usable catalogue of nebulae, is NGC 598 in the *New General Catalogue of Nebulae*, published by J. L. Dreyer in 1888. In his catalogue, Messier recorded slightly more than 100 apparently non-stellar objects without specifically considering their actual nature. As a result, Messier's *Catalogue of Nebulae* lists 29 open clusters, 29 globular clusters, 10 diffuse and planetary nebulae, and 34 galaxies. Dreyer's *New General Catalogue* contains similar objects, but has a total number of more than 6000.

The galaxy M33 (or NGC 598) is situated in Triangulum and has an apparent visual brightness of magnitude 5.8. Its distance amounts to 7×10^5 pc. In the sky, the galaxy covers an area 68 minutes of arc by 40 minutes of arc. From this, and its distance, a diameter of at least 13,000 pc can be derived. M33 is therefore considerably smaller and less massive than our own Galaxy.

The photographs show clearly that M33 is a spiral system. The spiral arms are wide open and by comparison with the nucleus are inconspicuous. M33 is therefore classed as an Sc spiral. In Hubble's classification of galaxies normal spiral systems are designated Sa, Sb or Sc spirals. In Sa spirals the nucleus predominates and the comparatively faint spiral arms are very tightly wound. In Sb spirals, to which our Galaxy belongs, the arms are farther open and the nucleus is less dominant. In Sc spirals the nucleus is completely overshadowed by the wide-open arms.

In normal spiral systems the arms leave the nucleus tangentially. There is another type known as barred spiral. In these systems, two short, straight arms arise at opposite sides of the nucleus and the main spiral arms sprout at right angles from the ends of the bar thus formed. These SB systems are subdivided in exactly the same way as the ordinary spirals, i.e. as SBa, SBb, or SBc types, using precisely the same criteria.

The majority of all observed extragalactic stellar systems are spiral systems. Elliptical systems also exist, and these show no spiral arms, but only a fairly compact, extended area of luminosity. If such a galaxy appears circular, it is described as an E0 system. Elliptical galaxies are classified from E1 to E7, according to the ratio between their longest and shortest axes. An E7 system is the most elongated; in other words, it has the greatest ratio between the dimensions of the two axes.

Spiral and elliptical systems have a regular structure. About 2–3% of all galaxies have no recognizable shape, however, and these are described as irregular systems.

Hubble's morphological classification scheme of 1926 has been modified slightly several times but has never lost its overall significance.

Looking at the three illustrations of M33, it is clear that the spiral arms are more pronounced in one picture and less obvious in one of the others. The reason for this is that the exposures were taken at different wavebands. The picture in which the nucleus is clearest and is surrounded by a diffuse, luminous region, and where only traces of the spiral arms can be seen was taken in the red with a standard wavelength of 720 nm. In contrast to this, the other half-page illustration shows the spiral arms very clearly and they are still distinctly visible a great distance away from the galactic nucleus. This picture was taken in the ultraviolet with a standard wavelength of 350 nm. The whole-page photograph is an exposure taken in the blue region with a standard wavelength of 435 nm. Here the nucleus and spiral arms appear with comparable intensity. When compared with the exposures in U and R, the exposure in B has a fainter limiting magnitude, and so the whole system appears brighter. In comparing the three photographs, however, it is the ratio of nucleus and spiral arms in the different wavelength regions that is all-important. By comparing exposures taken in the short-wavelength U region and in the long-wavelength R region, we can conclude that the nucleus and the spiral arms consist of completely different objects, because one radiates predominantly in the red and the other predominantly in the blue.

Before pursuing this line of thought any further, it should be remembered that black and white exposures were used to obtain colour information. For this photographic plates are used that are only sensitive to a limited spectral range. In addition, colour filters are used that only transmit radiation at certain wavelengths. By a suitable combination of filter and plate just red or blue radiation can be selected, i.e. one black-and-white illustration contains information about just the blue part of the spectrum and the other about just the red part.

From physical investigations, it has been known for a long time that the colour at which the source radiates is dependent on its temperature. The higher the temperature, the greater the flux of energy in the radiation, i.e. the shorter the wavelength of the radiation. Because blue radiation has a shorter wavelength than red radiation, this means that the sources found in the spiral arms of M33 are hotter than those found in the nucleus. In general, the radiation that we receive from a galaxy is the sum of the radiation from all the stars contained in that galaxy. As a result, the observational findings about M33, as illustrated by the three pictures, indicate that the spiral arms are dominated by hot, blue stars, and

Fig. 6.4. The Triangulum Galaxy (M33) in Triangulum on a photo taken in the blue.

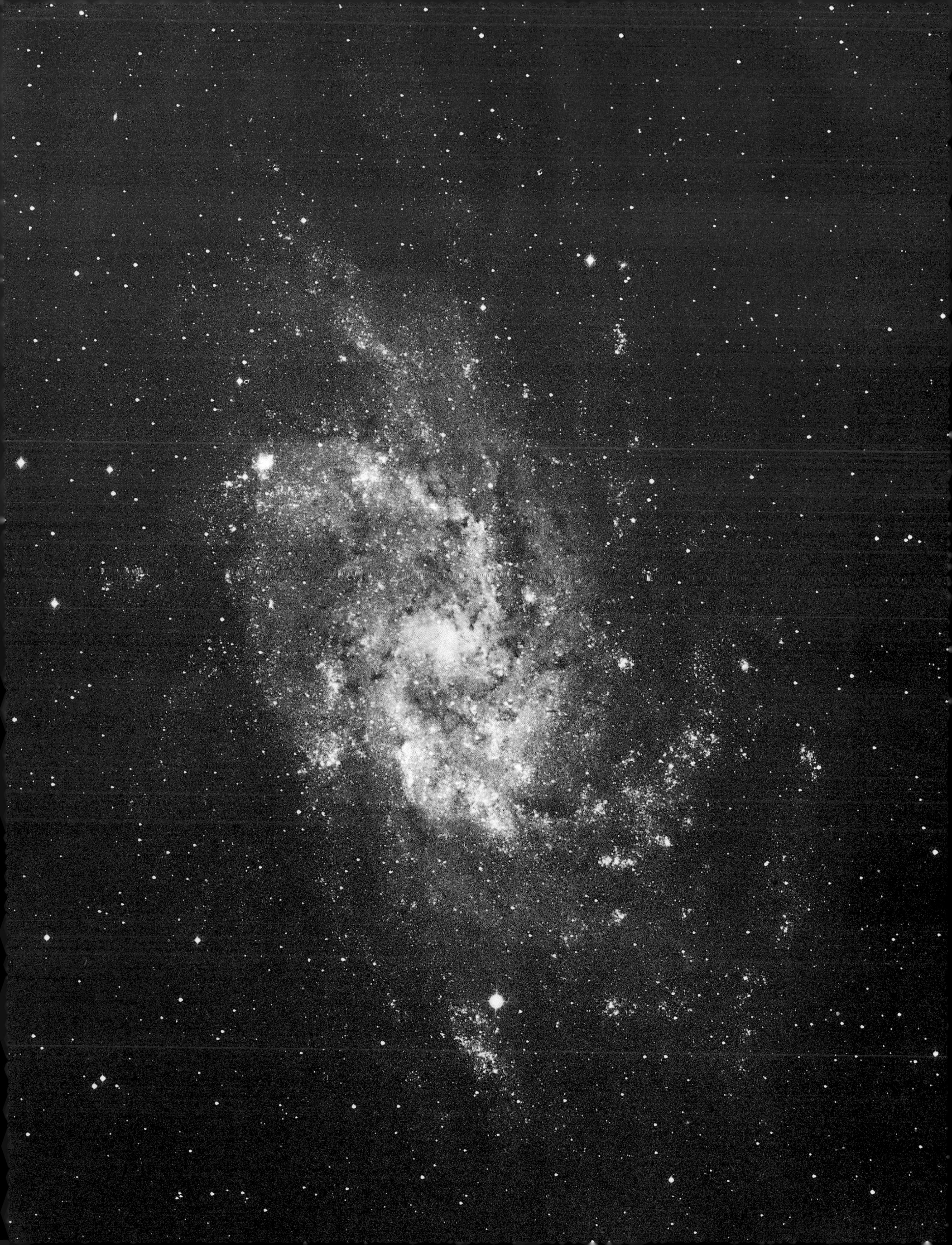

that the nucleus contains mainly cooler, red stars. This observation is not just typical of M33, but also applies to all spiral systems, and therefore to our Milky Way as well.

On the basis of this and other observational findings, Walter Baade introduced the concept of stellar populations in 1944. He distinguished two basic populations: Population I consists of objects in the spiral arms and Population II those in the nucleus of stellar systems as well as the objects that form a large, spherical halo around the flat, disc-shaped spiral system. Population I stars are younger than those of Population II. The population system was later improved, and stars are now grouped into five populations. The youngest stars are massive and have a high surface temperature. Their peak radiation emission is in the short-wavelength, blue spectral region. Open clusters and interstellar matter are also counted as members of this population, known as extreme Population I.

At the opposite extreme, as regards age, are objects of extreme Population II, such as members of globular clusters and certain RR-Lyrae variables. Between these two extremes there are two sub-groups as well as the disc population. Novae and planetary nebulae are conspicuous members of the latter, but it also includes the majority of stars in the Galaxy.

The populations differ in chemical composition as well as in spatial distribution and age. Population II stars, for example, contain only 0.3% of elements heavier than helium, whereas those in the disc population have up to 2%, and in those of extreme Population I as much as 4%.

This example of pictures of M33 at different wavelengths illustrates how it is possible to obtain information about physical parameters and the structure of stellar systems with this special method of observation.

Fig. 6.5/6.6. Photos of M33 in Triangulum taken in the red spectral region (above) and in the ultraviolet (below).

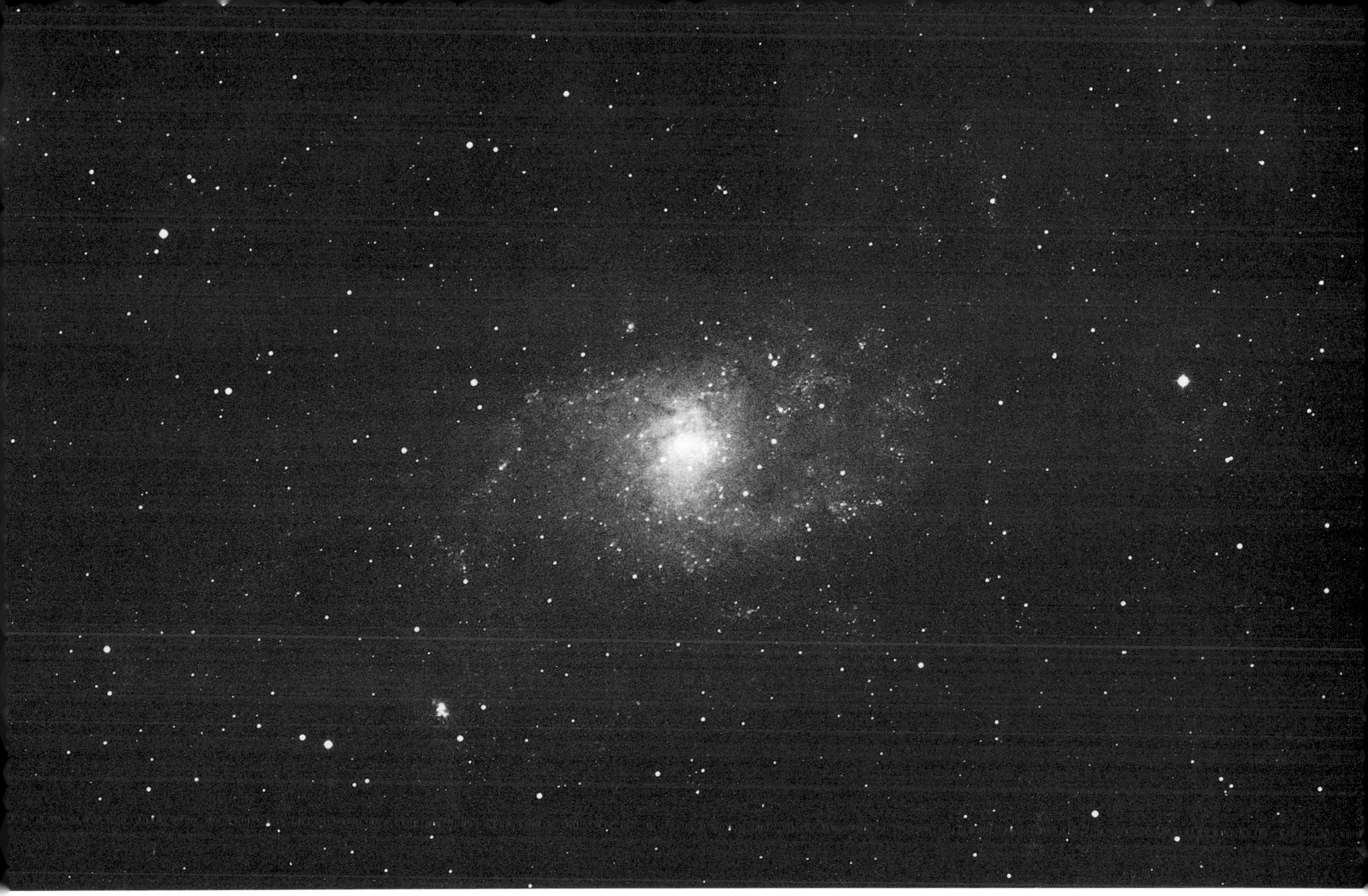

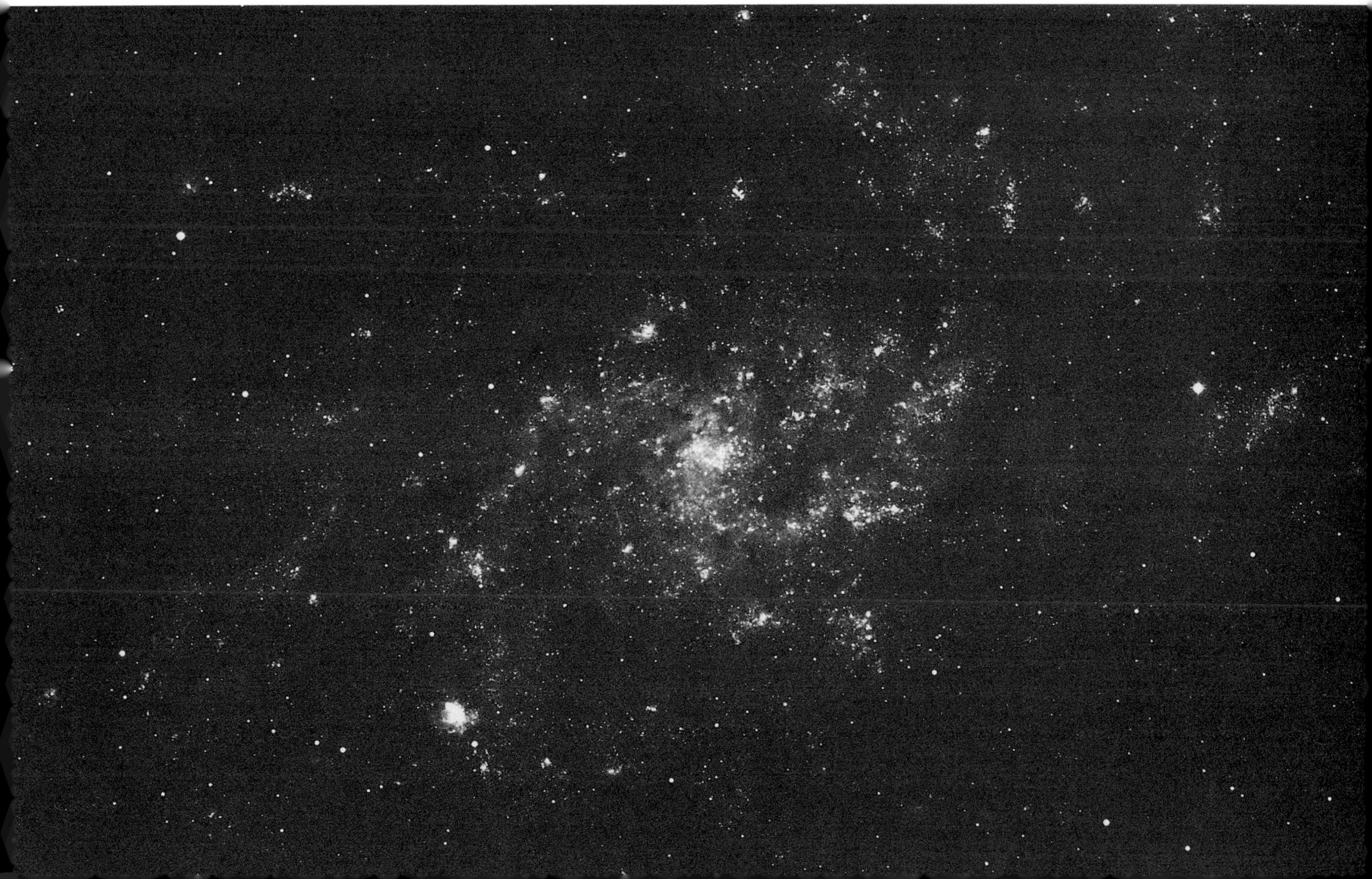

The open clusters h and χ Persei

It is possible to see two fuzzy objects in Perseus with the naked eye that are, in reality, clusters of stars. It is surprising that they are not included in the well-known Messier catalogue. They are both open clusters, and are described in the *New General Catalogue* as numbers NGC 869 (h Persei), and NGC 884 (χ Persei). These two clusters are among the most intensively investigated objects in the Galaxy. Nearly 500 scientific papers about each can be found in the literature. The first positional measurements of stars in h Persei were published as early as 1869, and the first scientific reports about χ Persei were written in 1885.

The angular distance between h and χ Persei on the sky is only 51 minutes of arc, most of which is in right ascension, because the difference in declination amounts to only two minutes of arc.

It is not by chance that these star clusters are to be observed in the same direction; they really are very close to each other in space. The distance of h Persei amounts to 1900 pc and that of χ Persei to 2000 pc. As both clusters have the same angular diameter of 30 minutes of arc, similar linear diametres of absent 17pc results.

About 300 stars are known in the open cluster h Persei and 250 in χ Persei. This corresponds to a spatial density of one star per cubic parsec, which is about six or seven times the stellar density in the vicinity of the Sun.

The clusters h and χ Persei contain stars of all spectral classes from O to M. From the existence of luminous, massive O-type stars we can conclude that NGC 869 and NGC 884 are still very young. Their age, only a few million years, is only about one-thousandth of that of the Sun and of the Earth.

In total, there are over 1000 open clusters known to exist in the Milky Way system. It can be assumed, however, that there are many more because large parts of the Galaxy cannot be seen at optical wavelengths, because of the effects of absorption by intervening interstellar dust. Like h and χ Persei all open clusters are relatively young members of our Galaxy and this can even be inferred from their very existence.

Open clusters are subject to various effects. Open clusters take part in the general rotation around the centre of the Galaxy. The stars in the cluster that are nearest to the galactic centre move more rapidly than those farthest from the centre, just as the orbital velocity of Mercury around the Sun is considerably greater than that of Pluto. As a result of these different rotational velocities, stars in a cluster gradually move farther away from each other. The cluster disperses, and the stars spread out into the general stellar population. The gravitational effects of neighbouring masses, especially nearby clusters and molecular clouds also tend 'to pull the cluster apart'. In addition, all stars have their own proper motions which over a sufficiently long time will cause the members of the cluster to disperse.

The members of the cluster are bound together by their gravitational forces. As the number of stars in open clusters is not very great, and they are loosely distributed, the gravitational forces that bind them together are unable to withstand the forces tending to disperse them, and after several hundred million years at most the majority of clusters have dispersed.

NGC 869 and NGC 884 have a low galactic latitude of only -4°. They are therefore situated in close proximity to the galactic plane. This is a general characteristic of all open clusters.

Accurate determination of the positions and (above all) of the distances of open clusters shows that they are found only in certain areas of the Galaxy. From their spatial distribution, the conclusion can be drawn that young open clusters form part of the Milky Way's spiral arms. We have the Orion, Perseus and Sagittarius arms, for example. Radio observations, especially of neutral hydrogen at a wavelength of 21 cm, have convincingly confirmed this spiral structure in the Galaxy. The radio observations also show quite clearly, however, that the spiral structure principally consists of interstellar matter. This, in turn, is the material from which new stars are born, so that in the Galaxy young objects can mainly be expected in the region of the spiral arms. This includes open clusters like h and χ Persei, which are components of the Perseus spiral arm.

Fig. 6.7. Galactic clusters h and χ in Perseus.

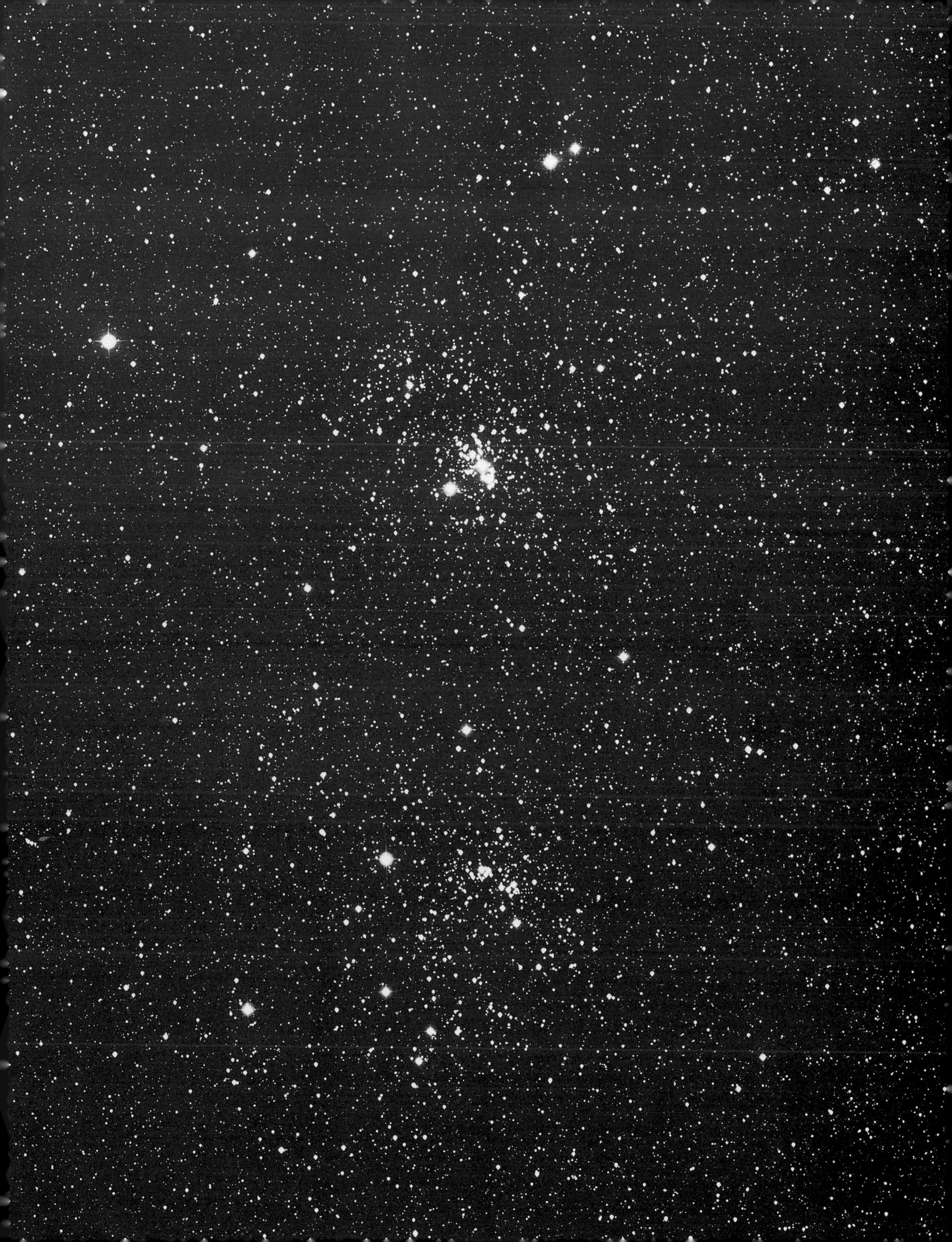

The Pleiades

A small cluster of stars can easily be seen in Taurus. Seven stars in this area have an apparent visual brightness of at least the sixth magnitude and under very good atmospheric conditions they can be seen with the naked eye. This cluster is therefore commonly known as the Seven Sisters.

Even with a simple pair of binoculars, it is immediately obvious that the number of stars with fainter apparent visual magnitudes increases considerably, and more rapidly in the immediate vicinity of the Seven Sisters than in the surrounding field. The photographs here were all taken with the same telescope. The different number of stars in the various pictures is a result of different exposure times. With a short exposure, only the radiation from the brightest stars leads to a darkening of the photographic plate. If the light is allowed to take effect for a longer time, then more fainter stars are recorded. Visual observations or simple photographic exposures reveal the Pleiades, as the Seven Sisters are also called, as a cluster of stars in the overall stellar field. With visual observation, only seven stars can be seen down to the sixth magnitude, but there are 230 stars in this area above magnitude 12. A telescope with an aperture of 12 cm is sufficient to show these. It is thought nowadays that the total number of stars in the cluster is about 1000. It is clear from the three exposures of the Pleiades taken with the Schmidt camera that not only an increasing number of stars is shown with increasing exposure times, but also that diffuse, luminous areas become visible, which cannot be seen with the naked eye or with small telescopes.

The Pleiades cluster is number 45 in Messier's catalogue of 1784. It is situated not far from the Milky Way, in which numerous other star clusters can be observed. The Pleiades, like the other open clusters in the Milky-Way region, has no regular structure, unlike globular clusters. The Pleiades are an open cluster.

The Pleiades attracted the attention of astronomers very early on. In 1841, Bessel measured the positions of the stars in the Pleiades very accurately using a heliometer. These measurements were repeated with similar precision by other astronomers and the new positions were compared with those of Bessel. As a result of this comparison, the 1914 edition of *Popular Astronomy* by F. N. Newcomb and R. Engelmann states: 'By comparing his heliometer measurements with those of Bessel Elkin discovered that out of 52 stars, included in both series of measurements, 45 had almost the same proper motion as Alcyone, and therefore form a physically connected stellar system.' Alcyone is the brightest star in the Pleiades cluster.

The stars of the Pleiades cluster cover a relatively large area of sky. This in itself leads to the conclusion that this physically connected group is not very far away from the Earth. Several investigations have revealed a distance of about 120 pc, not very far in comparison to the distance to the centre of the Milky Way system which amounts to almost 9000 pc. Angular dimensions and distance indicate that the linear diameter of the Pleiades is a few parsecs. From the distance and the observed apparent magnitudes, the absolute magnitudes of the stars can be easily calculated. Very high values result for many of the Pleiades stars. There are several Pleiades that have absolute magnitudes five or six magnitudes greater than the Sun. A star with an absolute magnitude six magnitudes greater than the Sun radiates 250 times as much energy in a given time, and one five magnitudes greater, 100 times as much energy. Stars that release and radiate a lot of energy do so on account of their large mass. It is also known that the greater the mass, the shorter the life of the star; i.e. massive stars with high absolute magnitudes have a short lifetime. Applying these considerations to the Pleiades means that it must be a young cluster because the massive, high-luminosity stars still exist. As well as the very bright stars, there are, of course, also many low-mass stars in the Pleiades, which have the same age as the massive stars. The lifetime of less-massive stars is, however, far greater, i.e. they will exist for a longer period of time.

The age of the Pleiades has been estimated at about 60 million years. On the cosmic time-scale, therefore, they really are youngsters. Our Sun, with an age of 5×10^9 years is almost 100 times as old, and the Milky Way has existed over twice as long as the Sun. There are, however, younger clusters than the Pleiades in the Milky Way. It is estimated, for example, that the open cluster NGC 663 is only 6 million years old, just one-tenth of the age of the Pleiades.

The high-luminosity, massive stars radiate most energy in the short-wave, blue, spectral region. This implies that the surface temperature of these stars is very high. It lies between 15,000 and 25,000 K. Our Sun has a surface temperature of just 6000 K. Its radiation therefore peaks in the yellow spectral region.

On long-exposure photographs of the Pleiades, diffuse, luminous matter is visible over the entire region, but especially around the bright stars. We know that there is dust and gas in space between the stars. Spectral observations show just what sort of matter it is. The emission from the gas is dominated by a line spectrum, corresponding to its chemical composition. Dust, however, simply reflects starlight. This means that the spectral composition of

Fig. 6.8/6.9. Photographs of the Pleiades galactic cluster taken with different exposure times.

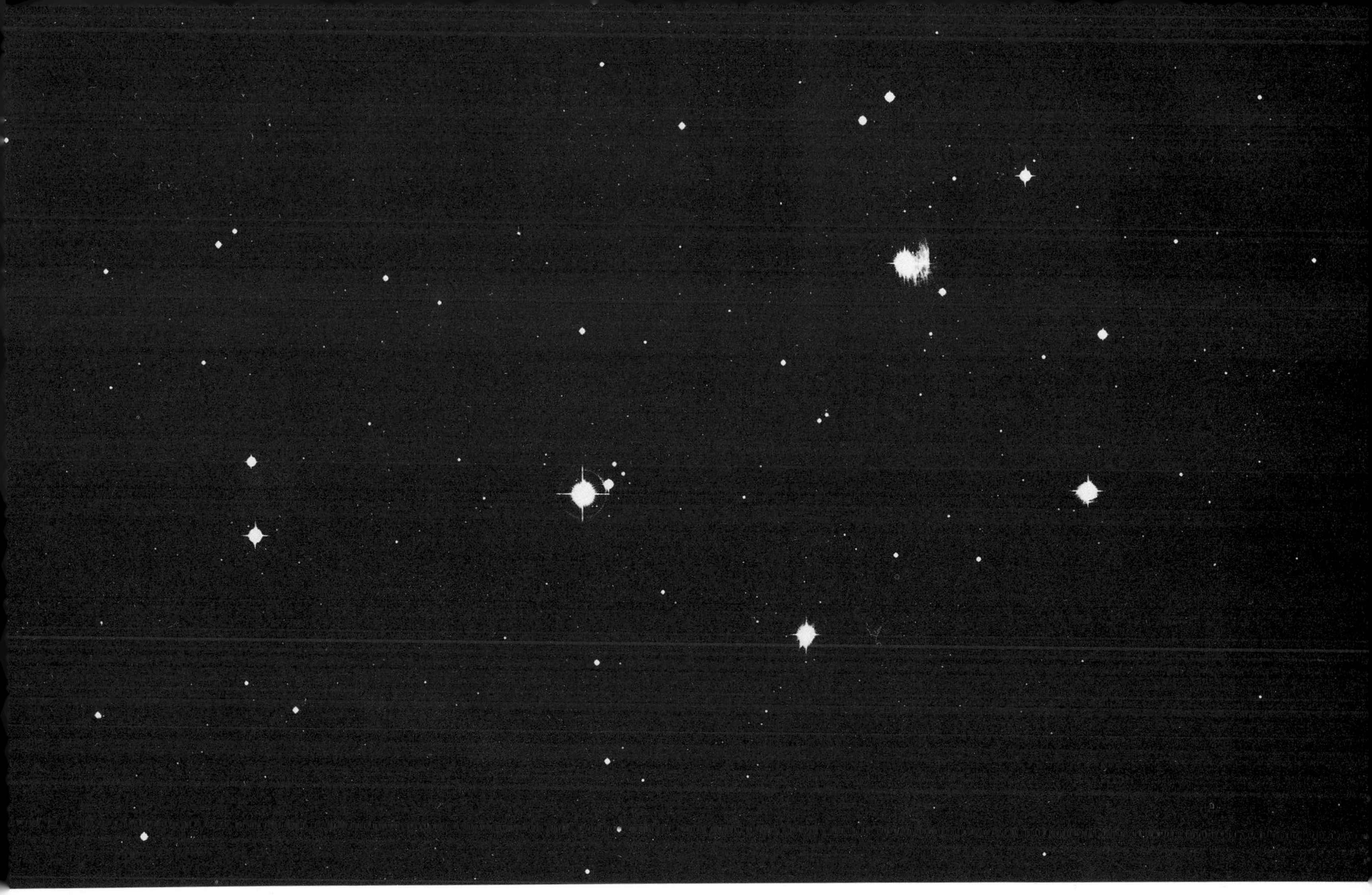

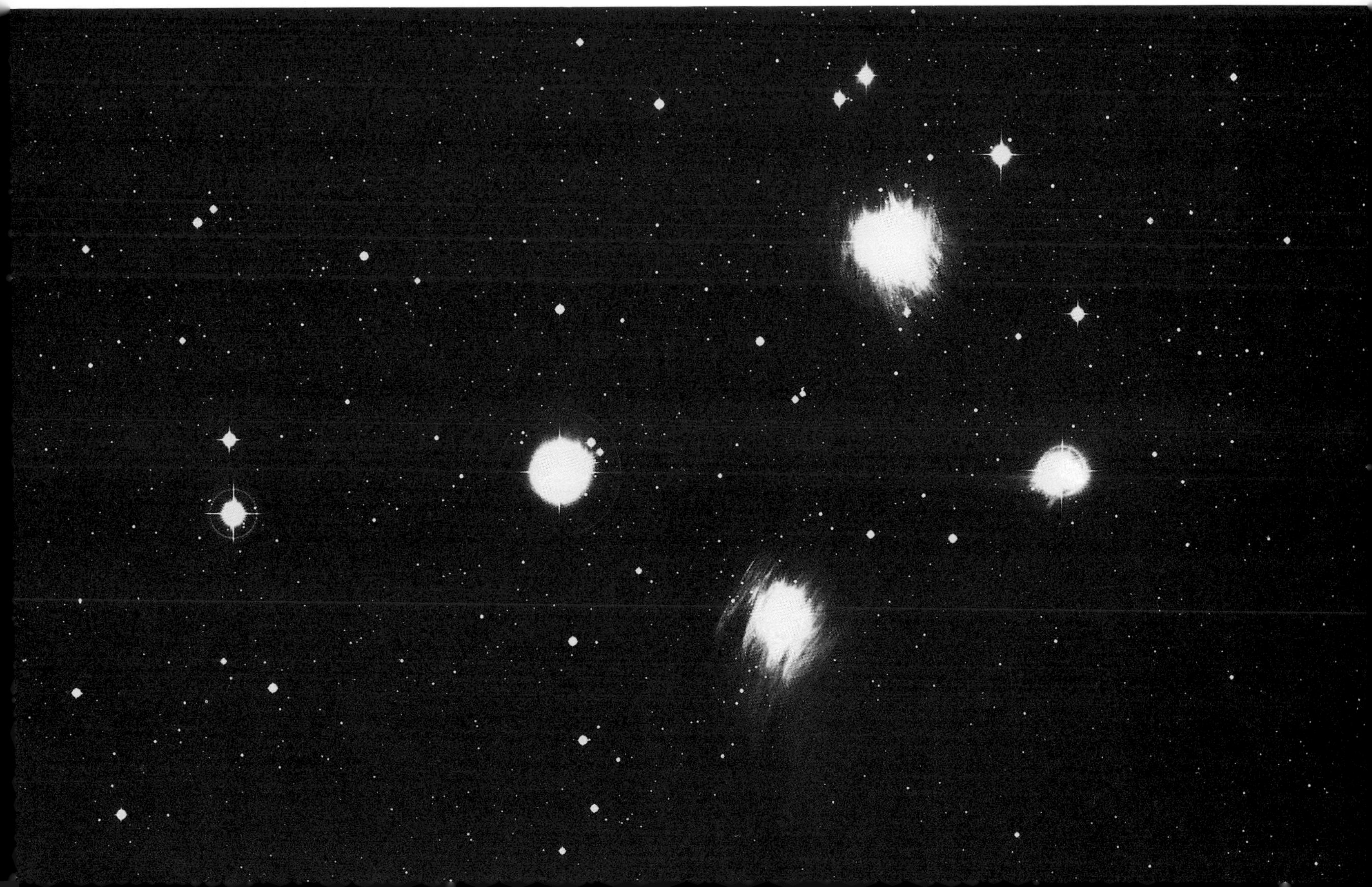

the light reflected by the interstellar dust particles corresponds to that of the stars themselves. In the case of the Pleiades, the radiation that we see is reflected by the dust component of the interstellar material. In the picture, the irregular distribution of the dust particles can be clearly recognized from the structure and the pattern formed by the swathes of dust.

The existence of interstellar material in the region of the Pleiades is directly connected with the age of this open cluster. We are certain today that stars form from interstellar material. For a star to form from interstellar matter a tremendous degree of contraction is necessary. The average density of the Sun amounts to about 1.4 g/cm^3 and the average density of an interstellar cloud is only about 10^{-23} g/cm^3. The mass of the Sun, which has a radius of 696,000 km, would have to be spread over a volume with a radius of about 1 pc in order to reduce its density to about 10^{-23} g/cm^3.

When a star forms by contraction from an interstellar cloud, not all the material is involved. Remnants of the original material are observed in the Pleiades cluster. Over the course of time, the diffuse matter becomes dispersed throughout space, and the stars themselves also move away from their birthplace. When stars are still young they remain closely associated with the interstellar material. With old stars such a close physical relationship is no longer observed.

The Pleiades and other open clusters also illustrate how stars are not formed as single individual objects, but that groups and clusters of stars always form together from very large interstellar clouds. The massive stars evolve very quickly and are the first to die, so that older open clusters increasingly contain only low-mass stars, which gradually disperse into space, until the cluster is eventually no longer recognizable.

The Pleiades are situated not far from the galactic plane, at a few degrees galactic latitude, and are therefore near the Galaxy's plane of symmetry. The other known open clusters in the Milky Way also lie in immediate proximity to the galactic plane, i.e. at very low galactic latitudes. More detailed investigation of the distribution of the open clusters shows that they are not evenly distributed along the plane of symmetry. The Milky Way is known to have a spiral structure, and the open clusters lie in exactly the same regions as the spiral arms and form an important source of information about this large-scale structure. Another indicator of spiral arms is interstellar matter. This again emphasizes the link between interstellar matter and open clusters, which are young components of the Galaxy. This means that in the Galaxy the area of the spiral arms are still zones of active star formation.

Fig. 6.10. The Pleiades galactic cluster in Taurus.

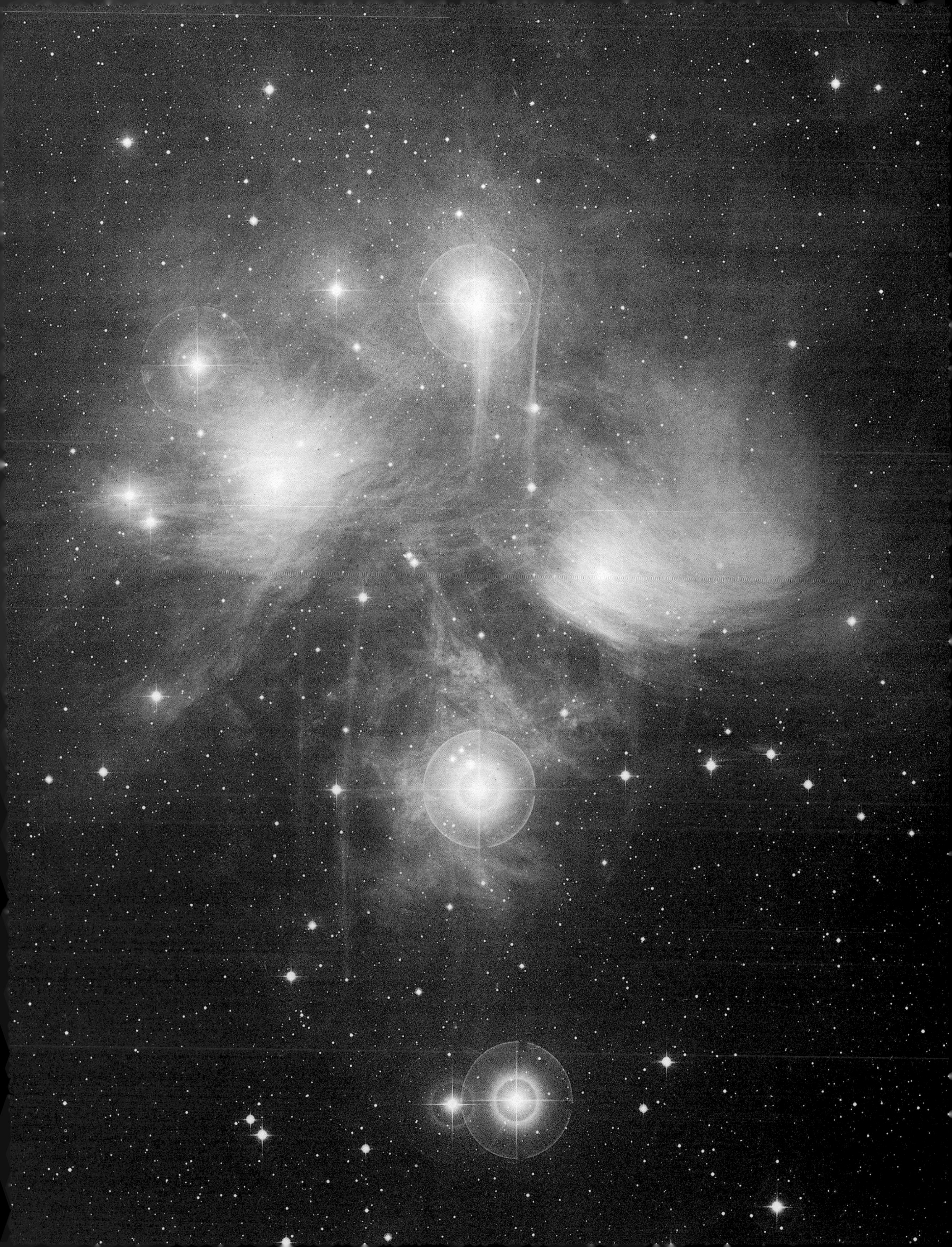

California Nebula

The vast spaces between the stars are populated by what, in terrestrial terms, is a very rarefied mixture of atoms and ions of gas. The gas density over large regions of the Galaxy amounts to only 0.1 atom per cubic centimetre. In fact, however, the matter is more typically distributed in the form of clouds, the denser regions of which may contain some ten, or even in extreme cases up to 10^5 particles per cubic centimetre. As in the universe as a whole, hydrogen is the predominant gas, with the addition of approximately 8% of helium atoms and trace amounts of atoms of elements with higher atomic numbers. Submicrometre-sized solid particles which make up 1% of the mass, are found in addition to the gaseous components. Under suitable physical conditions, both the gaseous and dust components of the cloud may appear luminous in the visible spectral region. Both the size and variety of forms shown by these objects mean that they are some of the most beautiful celestial phenomena. They are, however, very difficult to study because of their low surface brightness.

The nebula illustrated here is part of a cloud of luminous gas and is known as the California Nebula because of its apparent shape, which is reminiscent of the American state. The area that it covers on the sky amounts to some $2.5° \times 1°$, and its distance is 450 pc.

The star ξ Persei, which is the brightest star on the photograph, is responsible for the glow. Because it has an effective surface temperature of more than 35,000 K, the star emits a stream of about 10^{49} energetic photons per second into the gas in the nebula, which causes hydrogen atoms to become ionized over a large area. This forms what is known as an H II region.

In the equilibrium state recombination processes opposite to ionization also play a part, when protons and electrons that exist in the plasma combine. These processes are the source of the characteristic emission in the ultraviolet, optical, and infrared spectral regions, as well as at centimetre and decimetre radio wavelengths. This fact that we may make measurements over a whole range of the electromagnetic spectrum, which first became possible in the 1960s, is one of the foundations on which our knowledge of diffuse nebulae – which by now has become very extensive – is built. The other foundation is our theoretical understanding of the physical processes of excitation and emission of radiation in gaseous nebulae. This is based on quantum mechanics, which was developed in the 1920s, and has been continuously refined subsequently leading to further opportunities for research. Because of its capacity for covering an extended area, photography with Schmidt telescopes has also made an important contribution to our understanding of diffuse nebulae. It enables the large-scale distribution of luminosity over the sky to be studied in detail.

According to theory, a star embedded in a gaseous nebula should ionize hydrogen in a spherical region of space surrounding it. The radius of this Strömgren sphere – named after a Danish astrophysicist – of ionizing radiation characteristically depends on the star's parameters and on the density of the gas. Wide-field photographs show that in general, actual conditions differ considerably from the ideal case. Some reasons for this are variations in the spatial distribution of the gas and the screening effects of co-existing or intervening dust, and which may alter the amount of radiation, or may be seen as dark regions projected against the bright nebulosity. Both of these factors affect the appearance of the California Nebula.

In regions of ionized hydrogen, the electron density per cubic centimetre can be deduced from the strength of the continuum emission at radio frequencies. Bremsstrahlung radiation at wavelengths of a few centimetres, from electrons affected by the fields of positive ions in the plasma, is particularly important in this respect. The linear diameter of the emission region can be determined from a knowledge of the angular diameter of the Schmidt plates, and of the distance.

Combining electron density and diameter provides an astrophysically significant way of classifying H II regions. As density decreases and size increases, they range from ultra-compact and compact regions to dense, yet diffuse, objects, and, finally, to very diffuse, giant, H II regions. The optically visible California Nebula is part of a large H II region of the last type.

Fig. 6.11. The California Nebula in Perseus.

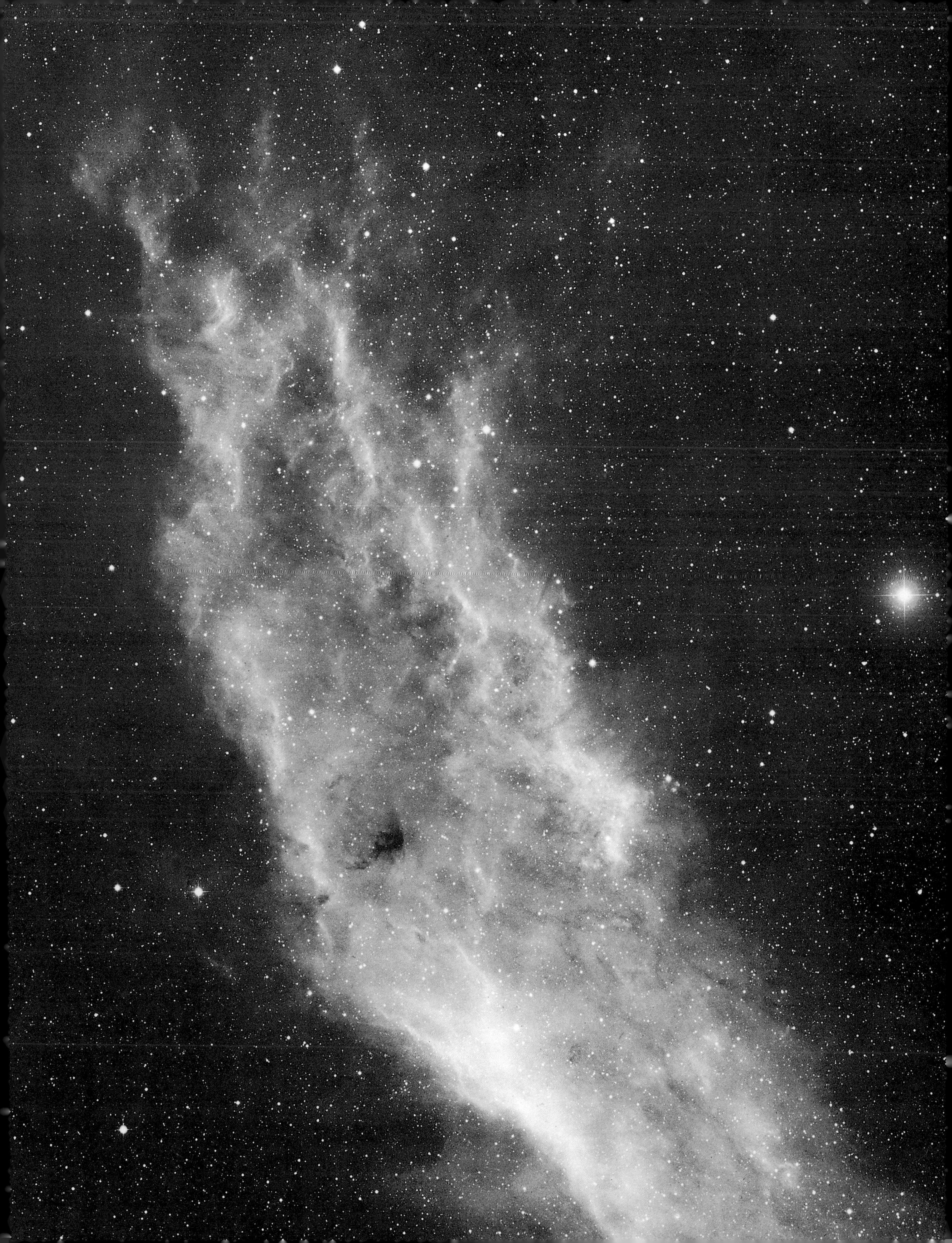

The Magellanic Clouds

Two of the most impressive sights in the southern sky are adjacent, extended, luminous objects, the first known descriptions of which were made in 1519 by Magellan, the Portuguese navigator. The distances determined for the Magellanic Clouds – as they were named after the circumnavigator – show that they do not belong to our Galaxy, but that they are independent stellar systems in space. Their distance from the Milky Way is slightly more than 50,000 pc, so they are the closest galaxies to our own system. Their appearance, however, differs greatly from the majority of known galaxies, which appear as spiral or elliptical systems. The Magellanic Clouds – as can clearly be seen from the pictures – have hardly any regular structure. They are therefore also frequently regarded as prototypes of the so-called 'irregular galaxies'.

At the end of February 1987, a supernova exploded in the Large Magellanic Cloud (LMC). It was given the designation 1987A. It was the first such event for 383 years to be visible to the naked eye. Its great luminosity allowed detailed investigations to be carried out in every wavelength region. The greatest sensation was the detection of a burst of neutrinos from the event. The scientific repercussions of many of the investigations that were carried out have still to be determined. Because of their proximity to the Galaxy, it is possible to resolve the two Magellanic Clouds into individual stars without much difficulty, as a fundamental step in determining their nature. The main objects that can be observed in the LMC are gaseous nebulae, dark absorption nebulae, supergiants and δ-Cephei stars. These are all members of Population I, and are therefore relatively young objects. Alongside these one also finds representatives of the disc population, and of Population II or the halo population. Novae, globular star clusters and RR-Lyrae variables, for example, all belong to the last population.

In the Small Magellanic Cloud (SMC) interstellar matter is not so prominent, but in comparison with our Galaxy, the ratio of interstellar to stellar matter in both the SMC and the LMC is still very high. More members of Population II can be found in the SMC than in the LMC, although young, Population I objects still predominate in the Small Cloud.

If the distribution of the different objects is studied in detail, some structure can be seen especially in the Large Magellanic Cloud. The important morphological characteristic is a stellar bar, as can be seen in the picture. In the Milky Way system, ionized-hydrogen regions – i.e. emission nebulae – are among the indicators of the spiral structure that is typical of our Galaxy. The distribution of the gaseous nebulae in the LMC also seems to suggest a weak spiral structure. For this reason the LMC is now frequently regarded as an extreme form of a barred spiral.

A quite noticeable difference between the Magellanic Clouds and the Galaxy, and many other stellar systems, is their very high content of interstellar matter. In the Galaxy this amounts to 4%, in the LMC over 10%, and in the SMC as much as 25%. The cause of this large amount of interstellar matter is probably a particularly slow process of star formation in the Magellanic Clouds.

The numerous young supergiants are the sources of the energy producing the many regions of ionized hydrogen, known as H II regions. The largest H II region in the Large Magellanic Cloud is the 30 Doradus Nebula otherwise known as the Tarantula Nebula. It has a diameter of 200 pc, and it contains one million solar masses of ionized hydrogen. A very large amount of energy is required to ionize a mass this size. Photometric observations have found over 100 stars with spectral classes earlier than B0, in the Tarantula Nebula. These are young stars with an effective temperature of at least 25,000 K and more than 8000 times the luminosity of the Sun.

The brightest and hottest source of energetic radiation in the Tarantula Nebula is the object R 136a. There are basically two models for this extraordinary source of radiation. It is either a very compact cluster of several extremely hot stars, which, because of the distance, can no longer be resolved, or it is a single object. This would be a true hypergiant of about 2000 solar masses, 60–80 solar radii, and an effective temperature of approximately 60,000 K. Whether it is a cluster or a single star, its luminosity amounts to some 30–60 million times that of the Sun. There are arguments for, and against, each possibility. Speckle-interferometry observations seem to decide in favour of the stellar-cluster argument and indicate that the object consists of three individual condensations. Exceptionally high quality photographic exposures also indicate bright condensations in R 136a. According to these results, R 136a is not a single object.

Evidence that it is a hypergiant is provided by the observation of a spectral line of carbon, which indicates a stellar wind of 3500 km/s. This is the highest velocity that has ever been measured for a stellar wind. If R 136a were a compact star cluster, then all the stars would have to have the same extremely high wind velocity. This is highly improbable. Optical spectroscopic observations show a Balmer-line redshift of approximately 100 km/s. This redshift can be interpreted as a gravitational redshift caused by a very massive object. The

Fig. 6.12. The Large Magellanic Cloud in Dorado in a negative print.

measured redshift is, however, rather too large for an object of 2000 solar masses. A final explanation remains to be discovered.

The mass of the LMC amounts to 10 thousand million solar masses, which is approximately one fifteenth the mass of our Galaxy. The SMC has a mass of only 2 thousand million solar masses.

The SMC looks very compact. The distances of more than 150 stars were recently very accurately determined. It was found that these stars had distances ranging between 45,000 and 75,000 pc, i.e. that the SMC has a depth of 30,000 pc in the line of sight. By contrast, its diameter, which we can see, is only 5000 pc. This very extensive system of stars and interstellar matter is distributed over far too great a volume for it to remain stable for a long time, especially under the gravitational influence of the Milky Way system. It has also been determined that the SMC is extending in the direction of our Galaxy with a velocity of 40 km/s.

From radio observations in the 21-cm region, the conclusion has even been drawn that the SMC consists of two separate galaxies, one behind the other in our line of sight. The radio observations show that these two concentrations of matter are approximately 6000 pc apart, and that this distance is increasing at a rate of 30 km/s. In this context, the two objects are often known as the SMC Remnant (in the foreground) and the Mini-Magellanic Cloud (in the background). From the observational data the conclusion can also be drawn that the SMC is in the irreversible process of breaking up.

The reason for the SMC being 'torn apart' could be a possible near collision of the SMC and the LMC about 200 million years ago; the process enhanced by the proximity of the massive Milky Way system.

The two Magellanic Clouds are often described as satellites of our Galaxy. This is justifiable, because they do orbit our Galaxy, taking one thousand million years to do so. Because of the proximity of the two Magellanic Clouds, individual objects can be resolved within them, and astronomers were able to observe different types of stars at a very early stage. Compared with their overall distances, the different parts of each Magellanic Cloud may be considered to be at the same distance from us. As a result, we may assume that differences in observed apparent magnitudes correspond to actual differences in absolute magnitude. This is particularly important in investigations that aim to determine the differences in the absolute magnitude of different types of stars. As early as 1912, Miss Henrietta Leavitt found the connection between the period of δ-Cephei variables and their absolute magnitude by observing stars of this type in the SMC.

The periods of δ-Cephei stars lie between 1 and 50 days, and approximately one-third of them have periods between 3 and 6 days. In general, the periods remain constant over long time-spans. According to the shape of the light-curve, these stars are divided into two groups: the true δ-Cephei variables with a steep rise to maximum and a slower decline; and the ζ-Geminorum stars, in which the rise and decline last the same amount of time.

In the δ-Cephei stars the size of the surface changes in parallel with the change in luminosity, i.e. these stars expand and contract. Because of the periodic change in their radius they are also known as pulsating stars. The change in radius amounts to 10% of the mean radius, which for different δ-Cephei stars lies between 7 and 150 solar radii. Observations have shown that the surface area of a star at maximum and minimum luminosity is the same, and therefore has nothing to do with the light variation. Spectral observations show that the temperature also changes as the light varies. At maximum it lies approximately 1000 K above that at minimum. Naturally, the spectral class also alters with the temperature. Because the luminosity of a star is strongly dependent on temperature, the change in temperature governs the variation in light.

The variation in magnitude, and thus the period of δ-Cephei stars, is easy to observe. From the relationship between period and absolute magnitude, one can determine the absolute luminosity. From the comparison between absolute and observed apparent magnitudes the distance can be calculated. The period–luminosity law of the δ-Cephei stars that was discovered by Miss Leavitt from the Magellanic Clouds is of fundamental significance for extragalactic astronomy, because it is possible to determine the distance of all galaxies in which δ-Cephei stars can be individually identified and recognized from their light curves. The observations made in the Magellanic Clouds were quite a decisive step in our knowledge of the universe.

Fig. 6.13. The Small Magellanic Cloud in Tucana.

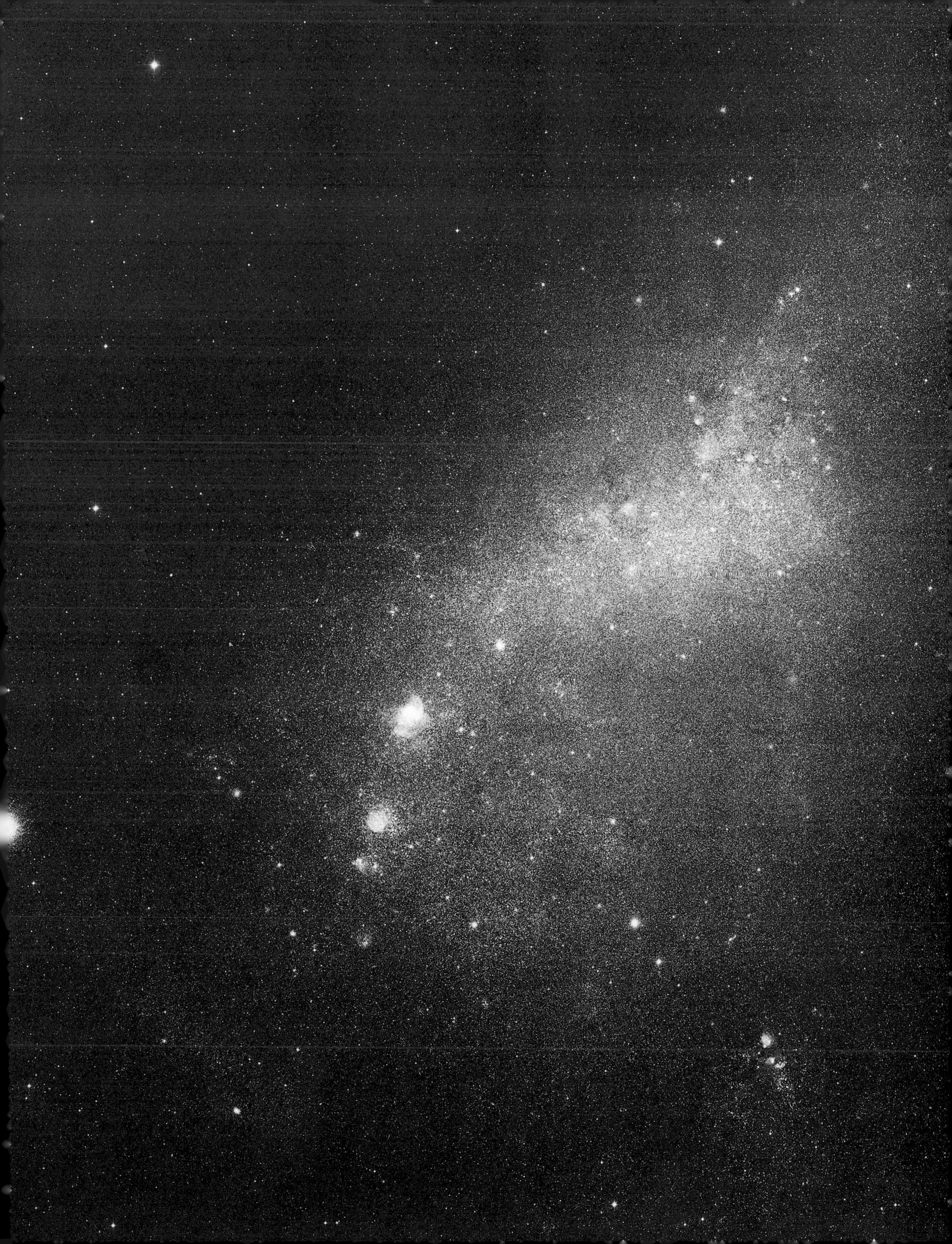

The Crab Nebula

The Crab Nebula is one of the most intensively investigated objects in the sky. This is because this emission nebula is exceptionally interesting astrophysically, and because of a particularly significant tale that lies behind its existence. The name is attributed to the Earl of Rosse, who made drawings of the object as seen through his giant telescope in the middle of the nineteenth century, and who described it as looking like a crab. A decisive discovery was made in 1921, when it was discovered, by comparing different photographic plates made at Mount Wilson Observatory, that in one decade the Crab Nebula had significantly increased in size. It was possible to calculate backwards from this expansion and determine that the gaseous nebula came into existence as the result of a known, historical, cosmic catastrophe – the explosion of a supernova. In 1054, a Chinese court astronomer gave a detailed description of the appearance of a 'guest star' in the constellation of Taurus. The star appeared so bright that it was visible for over 20 days in daylight, and for nearly two years at night. This 1054 supernova is the second of only five, historically authenticated, events of this type known in our Galaxy. As there is a lack of any reliable methods of determining the Crab Nebula's distance, this remains rather uncertain. The values derived by various authors lie between 1000 and 2000 pc.

The spectrum of the Crab Nebula resembles that of a planetary nebula in that it shows emission lines of ionized oxygen, nitrogen and other elements. It differs from planetaries, however, in having an intense continuum component as well. The Crab Nebula shows different structure in photographs exposed at different wavelengths. The filaments are particularly clear in the long-wave spectral region, because they consist of ionized hydrogen, and radiate very strongly in $H\alpha$, i.e. at a wavelength of 656 nm. In the short-wave region, however, diffuse emission is observed from the whole object.

The physical nature of the continuum component could be explained once the Crab Nebula had been recognized as one of the most intense sources of radiation at radio wavelengths. The shape of the spectrum of this radio emission indicated that the explanation was a radiation mechanism that was new to astronomy: synchrotron radiation. By this, we mean radiation, covering a wide range of wavelengths, that is produced by high-velocity electrons that are forced to follow spiral paths along the lines of a magnetic field. This process, known to physicists from work with atomic particle accelerators, now plays an important role in our understanding of cosmic processes. Synchroton radiation is strongly polarized, and, as a result, polarization measurements of the Crab Nebula show a fairly uniform magnetic field, with a strength about one thousandth of that of the Earth. Compared with interstellar conditions this is a strong field.

The diffuse emission observed in the short-wave visible region can be directly related to synchrotron emission. This radiation, however, also causes ionization of the gas in the nebula, via the related $H\alpha$ emission, it is also indirectly responsible for the long-wave parts of the radiation.

The source of the fast electrons is the central pulsar, which, together with the gaseous shell that was ejected during the event, and which is now visible as the Crab Nebula itself, indicate the explosion of a once-massive star. The pulsar is a rapidly rotating, compact object of unimaginably high density, which emits regular pulses of radiation in a process that is still imperfectly understood. In the case of the Crab pulsar, these pulses have a frequency of about 30 Hz. The period is decreasing by about 0.036 milliseconds per day, as the star's rotational energy is being converted into radiation. The fact that this pulsar is not only identifiable in the radio, X-ray and γ-ray regions, but also optically as a star, makes it astronomically unique. On deep Schmidt plates, a strange jet of material was detected with a length of about 75 seconds of arc, emanating from the northern border of the Crab Nebula. It is actually a hollow cylinder 45 seconds of arc in diameter, which means that it is about 0.5 pc across. Jet phenomena are known in quite different types of astronomical objects, but their physical nature is only poorly understood. The orderly structure exhibited by the jet is in contrast to the turbulent appearance of the nebula. In addition, we still have to explain why the jet does not point directly at the pulsar, which by general agreement should be the energetic heart of the Crab Nebula.

Fig. 6.14. The Crab Nebula (M1) in Taurus.

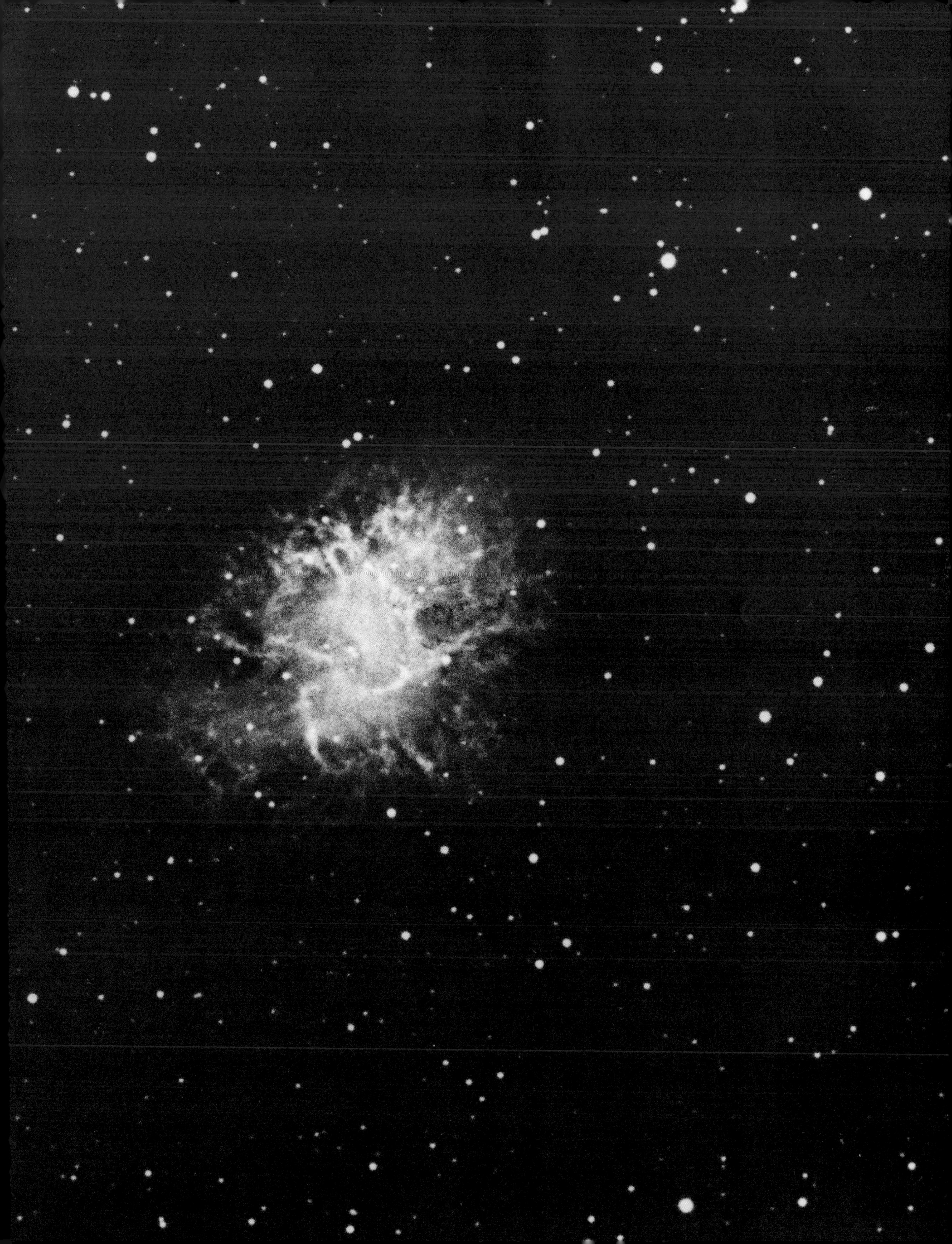

The Orion Nebula

A fine example of how impressive celestial objects can be is the Orion Nebula. Because this glowing nebula can be seen with the simplest of optical aids, its special role in the history of astronomical observation can easily be understood. Early drawings were made by Huygens with the aid of a telescope in the middle of the seventeenth century, and in 1880 it became the first nebula to be photographed. The fact that Huygens thought that the nebula was a hole in the sky through which a general, brighter background was shining gives us some insight into the way in which our ideas on the structure of the universe have developed.

In Messier's catalogue the extended nebulosity is designated M42. M43 is the smaller extension towards the north-east. To the north, and isolated from both, lies NGC 1977.

The fascinating filamentary structure of the Orion Nebula gives the immediate impression of an enormous, turbulent mass of gas. The luminous cloud of hydrogen and helium has a diameter of several parsecs, and lies at a distance of 450 pc from the Earth. The major part of the nebula's luminosity originates from the gaseous component, and in the visible region it mainly occurs in the form of discrete spectral lines. The sources of the luminosity are the four hot members of the multiple system θ^1 Orionis, which because of its shape is also known as the Trapezium. The neighbouring θ^2 Orionis system also contributes towards the luminosity of the nebula. Because of their high surface temperatures of between 25,000 and 40,000 K, the stars forming the source of the radiation pump an extremely high flux of energetic quanta into the surrounding gas, which has a mass amounting to about 700 solar masses. This ionizes the atoms in the gas, and the energy thus absorbed is converted into emission over a wide range of wavelengths.

θ^1 Orionis is a member of a very young cluster of stars that is 0.5 pc across, whose stellar density is more than 100 times higher than that of a typical galactic cluster. The fainter members of the cluster are still in the contraction phase, and this and certain other characteristics of the Trapezium system enable us to estimate the age of the nebula and its related stars as about 1,000,000 years.

The hot stars in the Orion Nebula form part of an even more extensive concentration of young, massive stars. These are distributed over the major part of the constellation, in what is known as an OB association, Orion OB1. Detailed analysis reveals that the entire association consists of distinct subgroups of differing size and age.

Apart from the western side, the whole area from the Trapezium as far as the central region of Orion is enclosed by 'Barnard's Loop'. This is an incomplete elliptical ring, measuring nearly 10° across, which appears as a faint extended object in the light of the Balmer Hα line and in the ultraviolet continuum. The 21-cm radio emission shows that the visible belt is surrounded by a zone of higher density neutral hydrogen and it is the inner edge of this zone that is ionized by the members of the Orion OB1 association. Barnard's Loop appears as the denser part of a much larger ring, which reaches into the neighbouring constellation of Eridanus. The entire complex may be evidence of a supernova event that occurred in this area a few million years ago.

Measurements in the infrared region at wavelengths below about 30 μm reveal among other things, a small area of increased emission within M42. This is the Kleinman–Low Nebula, which contains several sources of infrared emission. It is thought that this is a group of extremely young stars that are surrounded by a cloud of dense material and which are, therefore, not visible optically.

Associated with the Orion Nebula are two molecular clouds designated OMC 1 and OMC 2. OMC 1 lies in the direction of the boundary between M42 and M43; OMC 2 is near NGC 1977. The density of particles in these clouds is about 5×10^4 cm^{-3}, and the hydrogen appears in its molecular form, H_2. The presence of numerous types of molecules is a characteristic of these massive objects of dense, non-stellar material.

The combined data gives us a very good idea of the structure of this complex associated with the Orion Nebula. This data indicates that the luminous nebula and the stars in the Trapezium are part of the outer surface of the OMC 1 cloud that is closest to us. The gas surrounding the stars has been ionized by them and has broken through the front surface of the molecular cloud as a result of the increased pressure, streaming out to form the visible Orion Nebula. The stars of the Kleinman–Low Nebula are embedded in the molecular cloud. They are not at as advanced an evolutionary stage as the stars in the Trapezium. In the next 100,000 to 1,000,000 years, however, when the latter are no longer a strong source of ionization, the group of stars in the Kleinman–Low Nebula could well become the central energy source for a new gaseous nebula in this region. The process of star formation and the luminous gas that is associated with young, massive stars, so to speak, 'burning' into the molecular-cloud complex.

Fig. 6.15. The Orion Nebula.

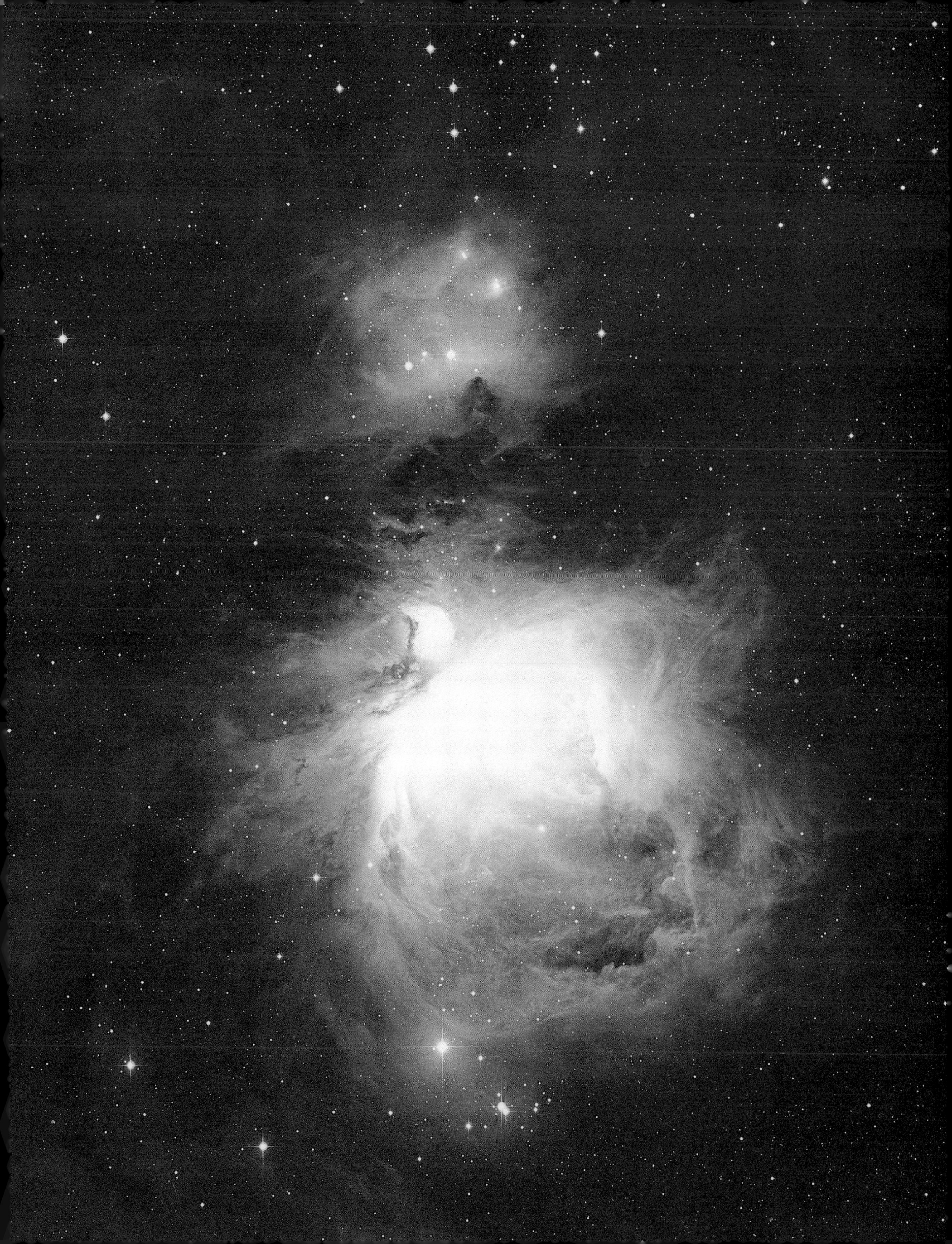

The Horsehead Nebula

This picture of the Horsehead Nebula shows several typical celestial objects. In one half of the picture the number of stars is clearly lower than in the other half. Apart from the areas that are rich or poor in stars, there are clearly defined, extended, bright regions. The most striking object is a dark area, devoid of stars, which gives the whole complex its name.

The Horsehead Nebula lies in the constellation of Orion, near ζ Orionis, a star in Orion's belt. This area shows a variety of forms of interstellar matter. The different number of stars in the two halves of the picture is caused by the scattering effect of intervening interstellar material. In the area where few stars are to be seen there is a large cloud of interstellar dust particles at a distance of about 300 pc. The dust particles are very small, generally with diameters between 0.01 μm and 1 μm, the smaller ones being most abundant. We know that most of the particles consist of silicates, often covered in mantles of frozen gases. This type of interstellar matter also includes solid carbon particles.

The spatial density of the interstellar dust is low, and within the clouds amounts to a maximum of 10^{-25} g/cm^3, corresponding to one dust particle in a cube of side 50 m. Dust particles at such a low density can still produce a distinct effect because of the gigantic volumes that are filled with dust particles, and which have diameters of several parsecs. With a density of 10^{-25} g/cm^3 and a cloud 3 pc in diameter, a total of some 2×10^{15} particles lie along the line of sight and contribute to the obsuration of the stars that lie behind the cloud.

Interstellar dust influences radiation through two different physical events. Particles of the size mentioned above cause scattering. This means that a fraction of the radiation is scattered from the line of sight, so that the flux received by the observer is less than it would have been in the absence of dust particles. Apart from this, there is true absorption of the radiation by the particles, resulting in a slight increase in their temperature, and a corresponding emission of radiation at longer wavelengths. This contribution is of minor importance in the optical region.

Both effects are dependent on wavelength. The short-wave regions of the spectrum are most strongly scattered. This means that the blue wavelengths are weakened when compared with the red wavelengths, and the light from stars that lie behind dust clouds appears reddened.

In astronomy, one talks about interstellar extinction, which combines the effects of absorption and scattering. The strength of the extinction depends, of course, on the spatial density of the dust particles and the extent of the dust cloud. The area in this picture that is poor in stars, and the Horsehead in particular, are explained in this way.

In the area poor in stars we can see two small, extended luminous areas not far from the Horsehead. These are known as reflection nebulae and are catalogued as IC 435 and NGC 2023. The glow arises because stars lie inside the dust cloud and are illuminating dust particles in their vicinity. These particles, in turn act to scatter the light, and the clouds of fine dust particles appear to glow. The spectral composition of the radiation from these reflection nebulae is essentially the same as that of the stars illuminating them. These nebulae, IC 435 and NGC 2023, are similar in nature to the well-known bright nebula surrounding the Pleiades.

The area of the Horsehead Nebula makes it perfectly clear that a single dust cloud may mimic an area poor in stars, but may also appear as extended luminosity if a bright star happens to lie within the dust cloud.

The Horsehead is seen in projection against a large area of luminosity which has its peak radiation in the red spectral range. Near the brightest star in the picture there is also a similar, strongly structured, extended area of luminosity. These extended regions are luminous interstellar hydrogen gas. This is excited by the short-wave, high-energy radiation from young, hot stars. The emission is particularly intense at the wavelength of the Hα line at 654 nanometers. The same physical process is involved as in the Orion Nebula, for example.

The area around the Horsehead Nebula shows in an impressive manner, and immediately adjacent to one another, the various ways in which interstellar material may appear: as luminous gas, as bright dust, or as dust absorption. This region is a portion of a very large complex of interstellar matter, to which the Orion Nebula also belongs.

Fig. 6.16. The Horsehead Nebula in Orion.

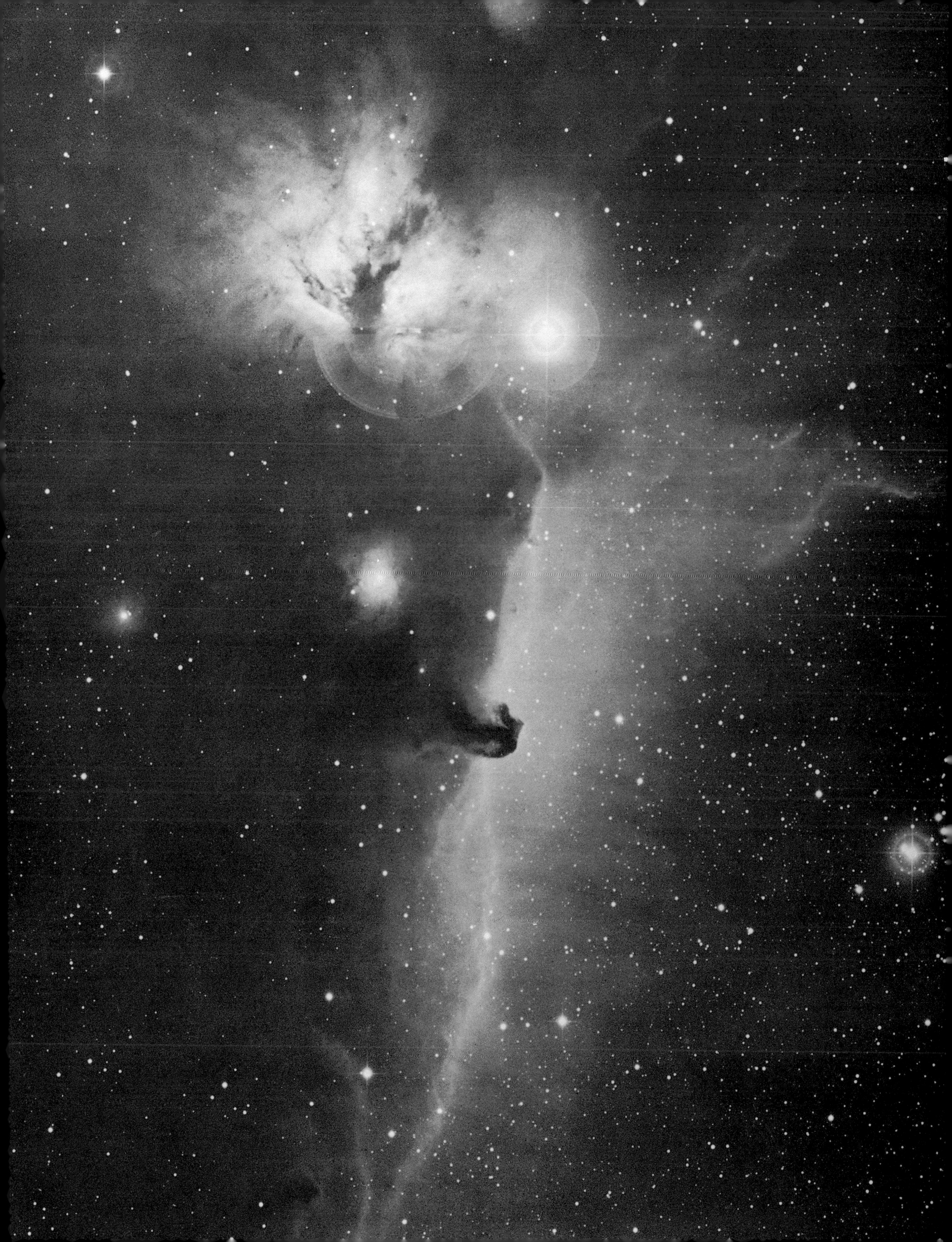

Open clusters

If a significantly large number of stars appears in one localized area of the overall field of stars, it may be a chance concentration, but it is more likely to be an open cluster. In contrast to globular clusters, they are not symmetrical, and even in the very centre of the cluster it is possible to resolve every individual star. As practically all open clusters are situated in the immediate vicinity of our Galaxy's plane of symmetry, i.e. close to the galactic equator, they are also known as galactic clusters. The name 'open cluster' refers to their structure; 'galactic clusters' to their position in the Galaxy.

Several open clusters can be seen with the naked eye, and therefore we may assume that they were known in ancient times. In fact the Pleiades and the Hyades were used as indicators of the seasons by farmers in Greece some 2700 years ago. Hipparchos, probably the greatest astronomer of ancient times, who, amongst other things, developed the theory of epicycles; recognized the eccentric motion of the Sun; and drew up the first star catalogue, referred to Praesepe, and h and χ Persei in about 150 BC. Because very few galactic clusters can be seen with the naked eye, scientific investigation of them began only after the invention of the telescope in the seventeenth century. The first phase in their study involved their being catalogued, and by 1920 about 150 objects had been listed. R. Trümpler, who made a major contribution to research concerning galactic clusters, catalogued a further 334 open clusters in 1930. The use of large astronomical telescopes led to the discovery of many more open clusters, and today about 1150 have been recorded.

It is not difficult to recognize that the appearance of an open cluster is closely linked to its distance from us. The angular diameter of a star cluster and the apparent distance between its stars get smaller and smaller the farther away the cluster is. This is made clear by this picture of two open clusters. The first cluster covers a very small area and the stars are very close together with a low apparent magnitude. The second has a large angular diameter, the stars are farther from one another and have a greater apparent magnitude. These simple indications tell us that the first cluster is farther away.

Trümpler developed a classification system for open clusters based on their overall appearance and the distribution of the stars within them. He defined four classes, which indicate the extent to which the clusters stand out against the general stellar background.

- Class I: The cluster is clearly distinguished from the general background of stars, and shows a strong concentration of stars towards its centre.
- Class II: The cluster is clearly distinguished from the background but shows only a weak concentration of stars.
- Class III: The cluster stands out against the background but shows no particular concentration of stars.
- Class IV: The cluster is hardly distinguishable against the background. It gives the impression of being a chance increase in the density of stars.

Further subgroups have since been added to these four main classes, and these subgroups take both magnitude and number of stars in the cluster into account. In each case, three subgroups have been defined. Those related to brightness are as follows:

- Group 1: The vast majority of stars in the cluster have approximately the same apparent magnitude.
- Group 2: All magnitude classes are found in the cluster, and are represented in approximately equal numbers.
- Group 3: There is a bimodal distribution: in other words, there are more bright stars and more faint stars than there are stars at intermediate magnitudes.

The three groups related to the number of stars are as follows:

- Group p (poor): Clusters in this group contain less than 50 stars.
- Group m (mean): Clusters with between 50 and 100 stars.
- Group r (rich): Clusters with more than 100 stars.

The classification of clusters into these last three subgroups is naturally influenced by their distance from the Earth. The farther away a cluster is, the fewer stars are visible.

Using these criteria, the classification of open clusters is now simple. Of the two clusters illustrated, the smaller cluster stands out more clearly against the general stellar background; has a concentration of stars towards the centre; and possesses a larger number of stars. It would therefore be placed in Class I or II, whereas the larger cluster would be placed in Class IV. With regard to the number of stars, the small cluster would be placed in group r, the other in group p. In the small cluster, all magnitude classes are almost equally represented; in the larger cluster, bright stars predominate.

The most famous cluster, the Pleiades, is classed as I3r. It is clearly distinguishable from the background; has a high concentration of stars; has both bright and faint stars but lacks those at intermediate magnitudes; and it contains more than 100 stars.

Fig. 6.17. The galactic clusters NGC 2158 and NCG 2168 (bottom left) in Gemini.

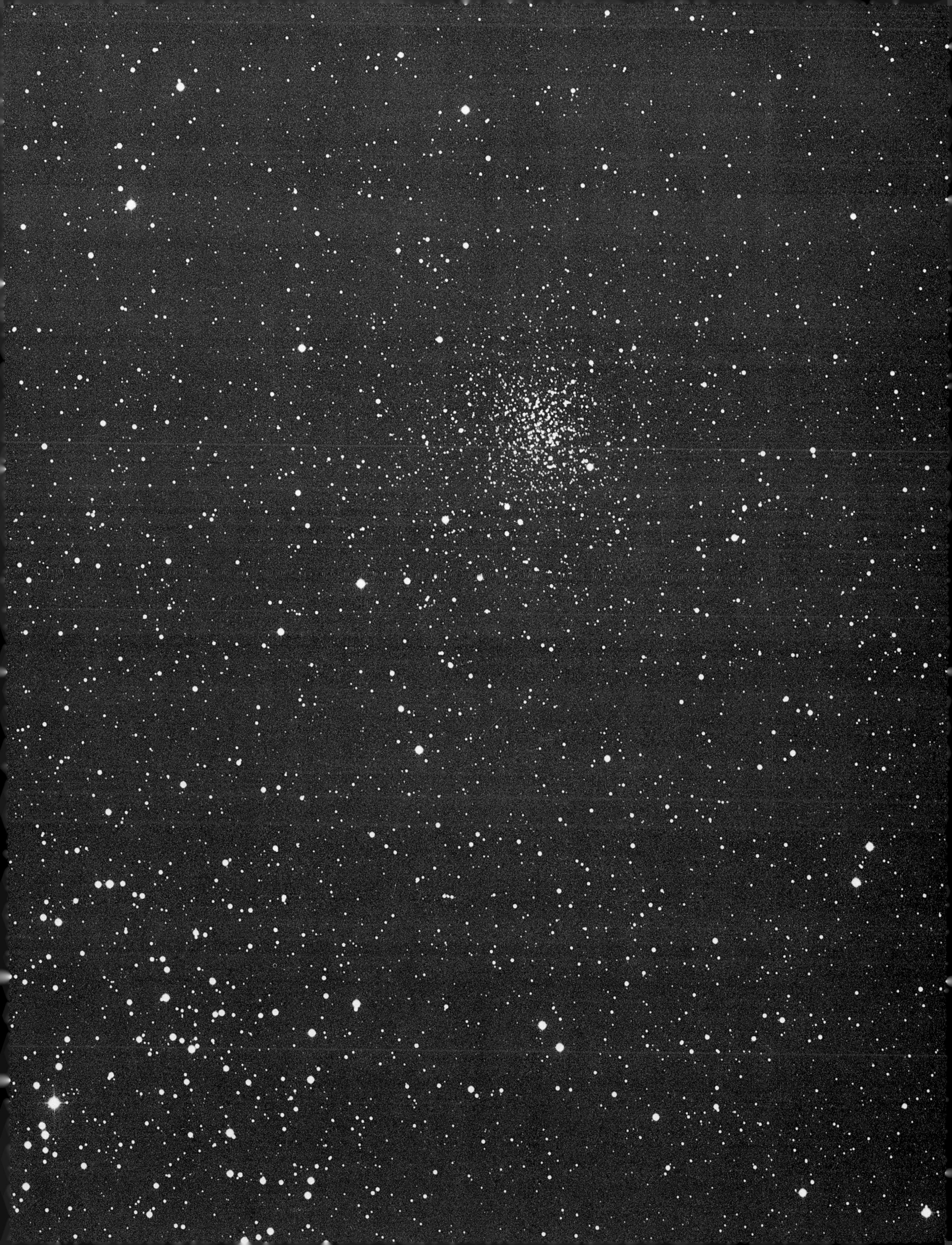

The Rosette Nebula

The symmetrical appearance of the Rosette Nebula, which has an angular diameter of about 1°, is caused by the spatial distribution of interstellar gas and hot stars. The nebula appears as a spherical shell, concentric with the young cluster NGC 2244. The latter can be seen in the central part of this photograph.

NGC 2244 is one of the youngest known clusters in our Galaxy. Its strong influence on its surroundings is caused by six stars of spectral class O, the hottest of which has a surface temperature of 45,000 K and a luminosity almost 10^6 times greater than the Sun's. This group of hot stars is responsible for ionizing the surrounding gas – which totals about 25,000 solar masses.

Investigation of this ring-like structure, visible in both optical and radio regions, confirms that the Rosette Nebula really does consist of a shell of ionized gas with outer and inner radii of 20 and 5 pc, respectively. The density of the particles in the shell amounts to about 15 cm^{-3}, but in the central cavity it is about one-tenth of this value. If the amount of ionizing radiation emitted by the stars is compared with the radio radiation emitted by the shell, we find that the stars are capable of ionizing exactly this volume of space. The external boundary is caused therefore by a lack of ionizing quanta, because more material is available, capable of appearing as an even larger nebula. Radio measurements of the 21-cm radiation from neutral hydrogen even show a slightly denser, outer H I zone.

The basic structure of the nebular shell is partly governed by the ionizing radiation, and partly by a continuous stream of particles emitted by the stars. This stellar wind originates in the outer layers of the stars. In the case of the hottest and most massive member of NGC 2244, mentioned above, this amounts to 2×10^{-6} solar masses per year. The stream of particles originating in the associated massive stars interacts with the surrounding plasma and thus produces the central cavity. In the course of the cluster's evolution, the rate of the mass-loss changes, as does the number of ionizing quanta. Calculations show that both quantities have greatly increased in the course of the last 300,000 years. This leads us to the conclusion that the structure of the nebula varies with time and therefore enables us to estimate its age. This proves to be approximately one million years, which is similar to the age estimated for the cluster from the stars' evolutionary stage.

Small dark 'globules' are frequently seen associated with luminous nebulae. These are very small, isolated dark clouds, normally regular in shape, which stand out against a bright background. Such globules are also found in the Rosette Nebula, especially in the north-western sector. Under closer examination they become more conspicuous because of their elongated shape and also because of their more-or-less radial alignment. At the boundary between the H II gas, which is expanding because of its higher pressure, and the neutral gas, various instabilities may form. It is then possible for tubes of neutral gas to form, projecting into the H II region, and these are frequently observed. Such features have been called 'elephants' trunks'. Such tubes may eventually be pinched off and compressed by the surrounding hot gas, thus forming globules. A larger type of globule is known in other celestial objects, having sizes of 0.5 pc and more, and these are regions where stars may form.

The cluster NGC 2244 and the surrounding Rosette Nebula lie at a distance of 1500 pc, and belong to the Monoceros OB2 association. This local concentration of stars of spectral classes O and B extends about 6° along the Milky Way. To the north, the Monoceros OB1 association surrounds the very young cluster NGC 2264, which has a similar age. Radio measurements show that extensive clouds of higher density neutral hydrogen and molecular clouds also form part of the same complex.

Fig. 6.18. The Rosette Nebula in Monoceros.

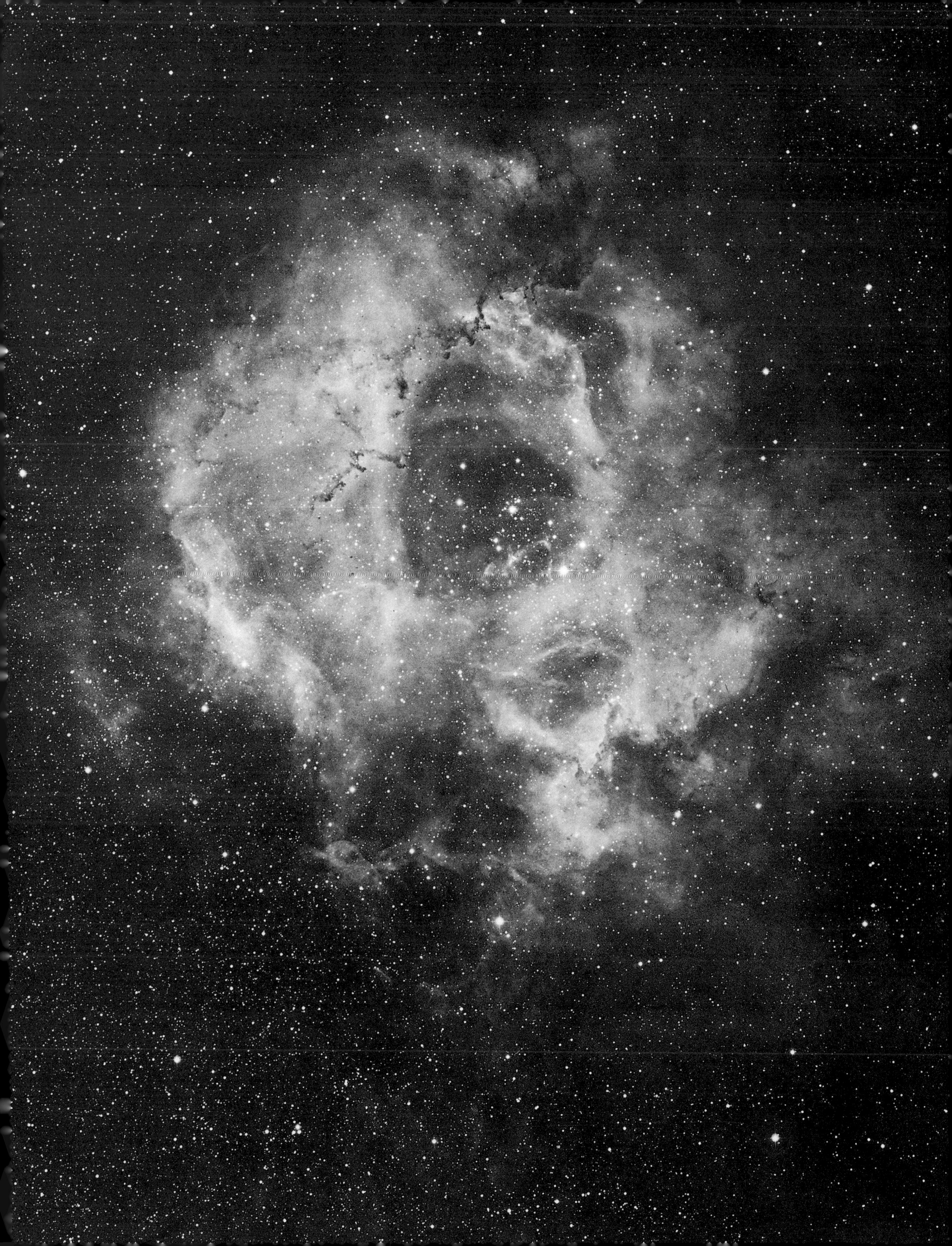

Hubble's Variable Nebula

Less than 1° away from the attractive Cone Nebula lies an inconspicuous comet-like nebula. This is Hubble's Variable Nebula, at the tip of which can be found the star R Monocerotis. This cannot be seen in the picture, however, because the nebula itself is too bright. The name, and the nebula's association with a variable star (also alluded to in the name), indicate that the nebula is of greater interest than its simple appearance would suggest.

NGC 2261 appears as a fan-shaped object, roughly $2' \times 1'$ in size, stretching out to the north of R Monocerotis. Based on its assumed distance of 800 pc, this means that its linear dimensions are a few tenths of a parsec. In 1916, Edwin Hubble drew attention to the nebula's variability in magnitude, shape and structure. High-resolution photographic plates, and visual observations using large telescopes under excellent atmospheric conditions, have since shown clearly that, although R Monocerotis does, indeed, look like a star, it appears quite diffuse when compared with other stellar images. In addition, the object is still seen to lie within a triangular nebula that is several seconds of arc in size. This is situated at the tip of the larger nebula, and is similarly fan-shaped, resembling NGC 2261, but with a higher surface brightness. Overall, modern observations support the view held by Lassell, an experienced observer in the last century, that R Monocerotis is a nebular knot rather than a star.

This variable nebula named after Hubble is by no means unique. A total of at least five other related objects have been recorded. They are all similar in their general appearance, in their relationship with certain types of variable star, and in their position inside, or on the edge of dark clouds. According to our current understanding they are part of a broader group of bipolar nebulae that have developed in the process of star formation. Knowledge about such early phases of stellar evolution has increased in recent years with the development of infrared astronomy and with research into interstellar molecules using radio astronomy.

The formation of a star begins when a portion of a dense interstellar cloud becomes unstable and contracts under its own gravity. As this occurs, rotation tends to form a nebular disc surrounding the protostar, which has already formed or is still in the process of formation. If the orientation is unfavourable, this disc can hide the star from sight. The thermal emission from such a disc can be identified in the infrared spectral region. The energy distribution in R Monocerotis fits in with this, because, apart from the normal decline towards longer wavelengths, it also shows another peak at 3.5 μm. This maximum indicates thermal emission from dust at a temperature of about 800 K. Speckle interferometry has shown that this dust shell spans about 1300 AU. In the plane of the disc, radiation at visible wavelengths is strongly absorbed, but it can escape along the rotational axis, and – rather like the cone of light from a searchlight – create a bipolar, dumbbell-shaped reflection nebula. There are examples where both of these lobes are visible, but also others, like NGC 2261, where one of the lobes is either invisible or scarcely detectable.

The instabilities associated with the final phases of star formation generally produce irregular fluctuations in the star's radiation, i.e. in its brightness. The nebula exists only because it reflects the light from the star, so we can expect a corresponding variation in its brightness. Observations have shown, however, that the intensity of light from the star and the nebula do not vary in parallel. There are variations in timing that suggest that although the star may appear to have one magnitude as seen from Earth, different parts of the nebula 'see' the star at very different magnitudes. This can be explained approximately by assuming that moving clumps of matter in the shell and in adjacent regions of the nebula tend to block light from the star. Detailed investigation of this explanation has taken account of the variable spectra of the star and of the nebula.

The discs around young stars not only cause these visible bipolar reflection nebulae, but are also responsible for the fact that the material driven away from the star by the stellar radiation does not expand isotropically into space, but instead, is strongly collimated along the axis of the disc. This produces radial streams of gas that have recently been intensively examined using various methods. These streams of material extend far out into surrounding space and remain evident as bipolar flows of molecular gas well away from the central object. The physical mechanism behind this strong focusing is not yet fully understood.

The objects known as Herbig–Haro objects are also connected with these long-range outflows. They are clumps of gas that appear star-like and which lie on the axis of symmetry of the internal motions of these bipolar sources. Collisions with particles are the cause of their emission.

Finally it should be mentioned that the particles of gas and dust that surround stars that are in the process of formation could be a preliminary stage in the formation of planets. The dimensions and mass of the discs are also compatible with such a hypothesis.

Fig. 6.19. Hubble's Variable Nebula in Monoceros.

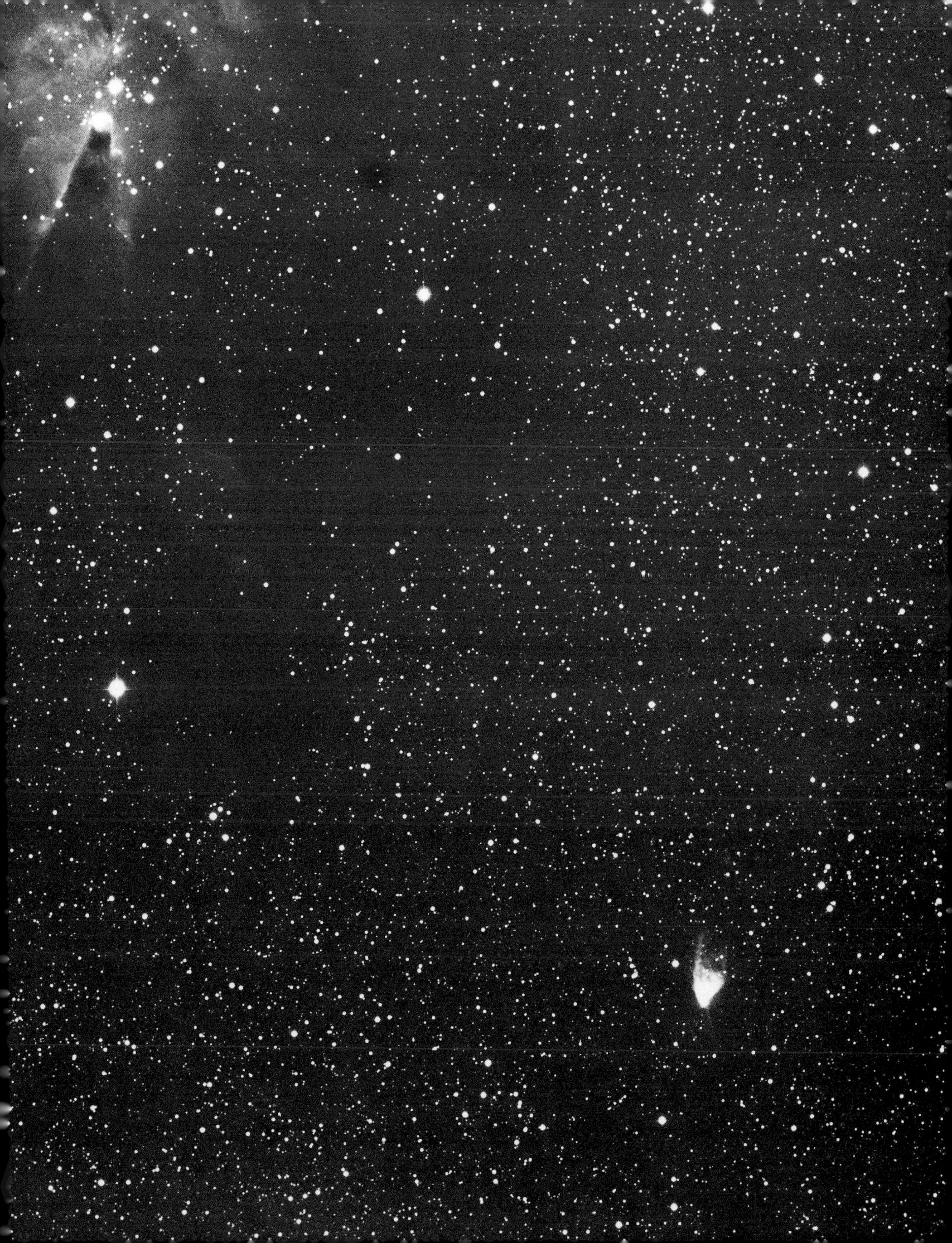

The Cone Nebula

The region of the Cone Nebula in Monoceros, which is visible in winter in the northern hemisphere, is one of the most interesting areas of the sky, because it shows many different forms of cosmic material in close proximity.

The most conspicuous object is the triangular cone, which is an area with very few stars. The cause of the lack of stars in the region of the cone, and in other similar areas of the picture, is the existence of interstellar dust, which scatters and absorbs light from the stars that lie behind the dust-clouds. The cone points towards large areas of luminosity that have a very chaotic structure. These generally conist of interstellar hydrogen, excited by hot, young stars. Sharply defined areas of absorption are seen in projection against these areas of luminous hydrogen.

The enlargement of the area showing the area immediately around the cone reveals that the bright point of the cone is an area of luminous interstellar material containing stars. These stars cannot be seen on the photograph of the whole area, because they are lost in the bright interstellar material. It is only because the characteristic curve has been flattened by a special photographic copying process that details can be detected here in the high-density areas. The appearance of stars at the point of the cone emphasizes the effectiveness of the process used. A comparison of the main exposure and the enlarged section also shows that not the slightest detail is lost in this process. In addition, the enlargement indicates the variety of structure present in the bright interstellar material.

The point of the cone also points to the brightest star in this area, which is known as S Monocerotis. Its apparent visual magnitude is 4.7, and can therefore be seen with the naked eye. S Monocerotis lies at a distance of nearly 1000 pc. The fact that it can be seen with the naked eye, despite its great distance, means that it must have a high absolute magnitude, i.e. it emits a lot of energy per unit time. If our Sun were removed to a distance of 1000 pc, it would be an inconspicuous star of magnitude 15, and observing it visually would require the use of a telescope with an aperture of at least 30 cm. The absolute visual magnitude of the Sun is 4.7, so S Monocerotis is about ten magnitudes brighter than the Sun. That means that the energy it emits must be nearly 10,000 times that of the Sun. S Monocerotis is a young star. It is not an isolated object, but is a member of the galactic cluster NGC 2264. A star is a sphere of gas at the centre of which energy is released by the fusion of atomic nuclei. The efficiency of this 'nuclear reactor' depends on the mass of the star, and determines its luminosity.

As long as 70 years ago, Ejnar Hertzsprung and Henry Norris Russell discovered that a relationship exists between the absolute magnitude of stars and their spectral class. The absolute magnitude is a measure of the star's luminosity and therefore provides evidence about the amount of energy emitted per unit time. The spectral class of stars is determined by their effective temperature, that is, the temperature of their surface. The effective temperature and luminosity of numerous stars can be determined and the values plotted on a temperature–luminosity diagram. It was soon realized that the diagram was not evenly populated by points, but that the overwhelming majority fall on a diagonal line in the diagram. On this line stars with high effective temperatures have high luminosities and stars with low effective temperatures have low luminosities. This line on the diagram, on which the values of nearly all stars fall is known as the main sequence.

When this procedure was applied to members of the galactic cluster NGC 2264, whose brightest star is S Monocerotis, it was found, with some surprise, that only the stars with the greatest luminosities lie on the main sequence, and the values for less luminous stars are clearly situated above the main sequence, the deviation increasing with decreasing luminosity.

This observation can be explained by the theory of stellar formation and evolution. We know that stars form from interstellar matter. The density of the interstellar matter is very considerably lower than the density of material in stars. The process of star formation is accompanied by an enormous compression of the material.

The force of gravity acts within an interstellar gas cloud, and this is dependent on the mass and radius of the cloud. Acting alone, gravity would cause the cloud to contract. In the interstellar cloud, however, there is also gas pressure which depends on the temperature, and which tends to force the cloud to expand. The force of gravity and the thermal pressure within the interstellar cloud therefore oppose one another.

Because, in a specific volume of space, gravitational attraction rises with increasing density, and gas pressure rises with increasing temperatures, the density has to be high and the temperature low in order for contraction to occur. Table 6.3 gives, for certain specific temperatures, the densities and the corresponding masses that lead to contraction.

The values of the first line are valid for interstellar space in regions between clouds of matter. Here, very large masses are required to trigger the contraction process, so that we need not consider regions between clouds as possible sites for star formation. Suitable conditions prevail, however, in high-density clouds. The mass of gas present there is too large for the formation of a single star, but fragmentation processes leading to the simultaneous

Fig. 6.20. The Cone Nebula in Monoceros.

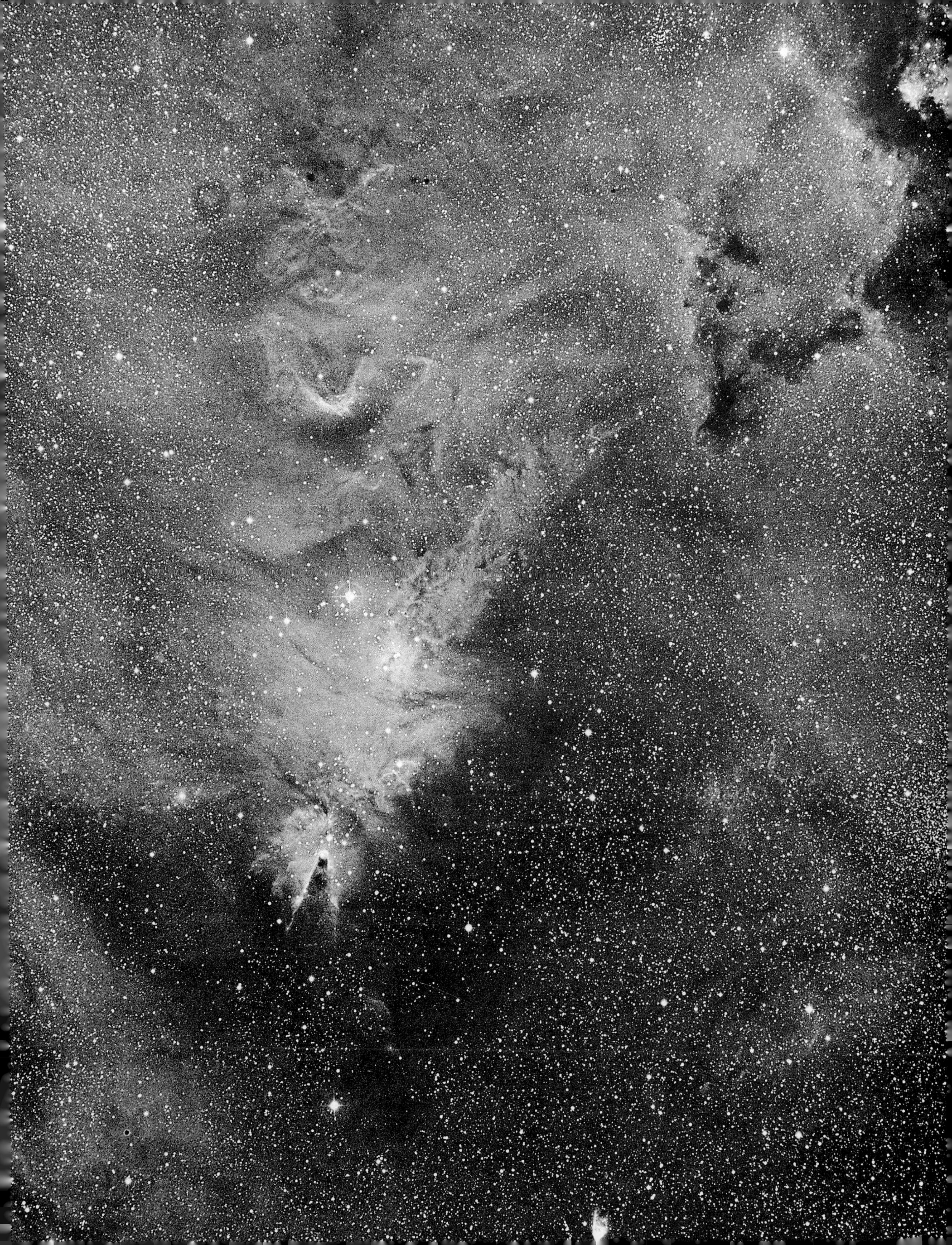

Table 6.3. *Minimum mass for contraction versus temperature and density*

Density (particles per cm^3)	Temperature (Kelvin)	Mass (solar masses)
0.2	8,000	50 million
100	100	3,000
1,000	10	30

formation of groups of stars may be expected to occur.

At a density of 100 atoms per cm^3 a cloud of 3000 solar masses is about 10 pc across. In slightly more than 10 million years the cloud contracts to a diameter of 3 pc. During this time, the gravitational energy released during contraction is mainly radiated away, so that there is no significant temperature rise in the cloud. As the contraction results in an increase in density, but not in temperature, smaller masses can contract separately. This begins a fragmentation process in the large cloud.

Within the objects created by fragmentation (or perhaps repeated fragmentation), density increases even further. During this process, such a high density is eventually reached that the energy released by the contraction can no longer be radiated away, and most of it is trapped in the object by absorption. This causes the temperature of the contracting object to rise, and it enters its first stable stage. The object is now described as a protostar, and the temperature continues to increase through contraction, until it reaches 10^6 K. At this temperature, the fusion of hydrogen to helium begins, which means that the protostar has become a stable star that is luminous in its own right. It is now an object on the main sequence in the temperature–luminosity diagram. The process of stellar birth is now complete.

The crucial physical process in the birth of a star is the contraction of the material. We can see the time-scale velocity of this process is determined by the mass of the protostar. The greater the mass, the higher the gravitational attraction and the faster the contraction. In the case of the Sun, the contraction phase lasted 30 million years; in the case of a star of only 0.6 solar masses it can last as much as 150 million years. Stars with 1.5 solar masses pass through this initial phase in 8 million years; and stars of 5 solar masses require less than 1 million years.

If these findings are applied to the galactic cluster NGC 2264, it becomes clear that NGC 2264 is a cluster in the process of formation. We may assume that the contraction phase began almost simultaneously for all the members of a cluster. The massive objects reach a stable, stellar state much earlier than the less-massive ones. This is precisely what is observed in NGC 2264. The high-luminosity massive stars, including S Monocerotis, already lie on the main sequence, but the low-mass stars are still in the process of formation and have not yet reached a stable, stellar state on the main sequence. Because of their high luminosity, massive stars have a very short lifetime, so they may complete their whole evolutionary cycle, and come to the end of their lives, before less massive stars have even finished the stellar-formation stage.

The region around the Cone Nebula can justifiably be described as an area of active star formation. Material suitable for the formation of new stars is abundantly available in the form of interstellar gas and dust, and in NGC 2264 we are currently observing just such a process of transition between interstellar matter and the stellar state.

Fig. 6.21. A negative print of the Cone Nebula.

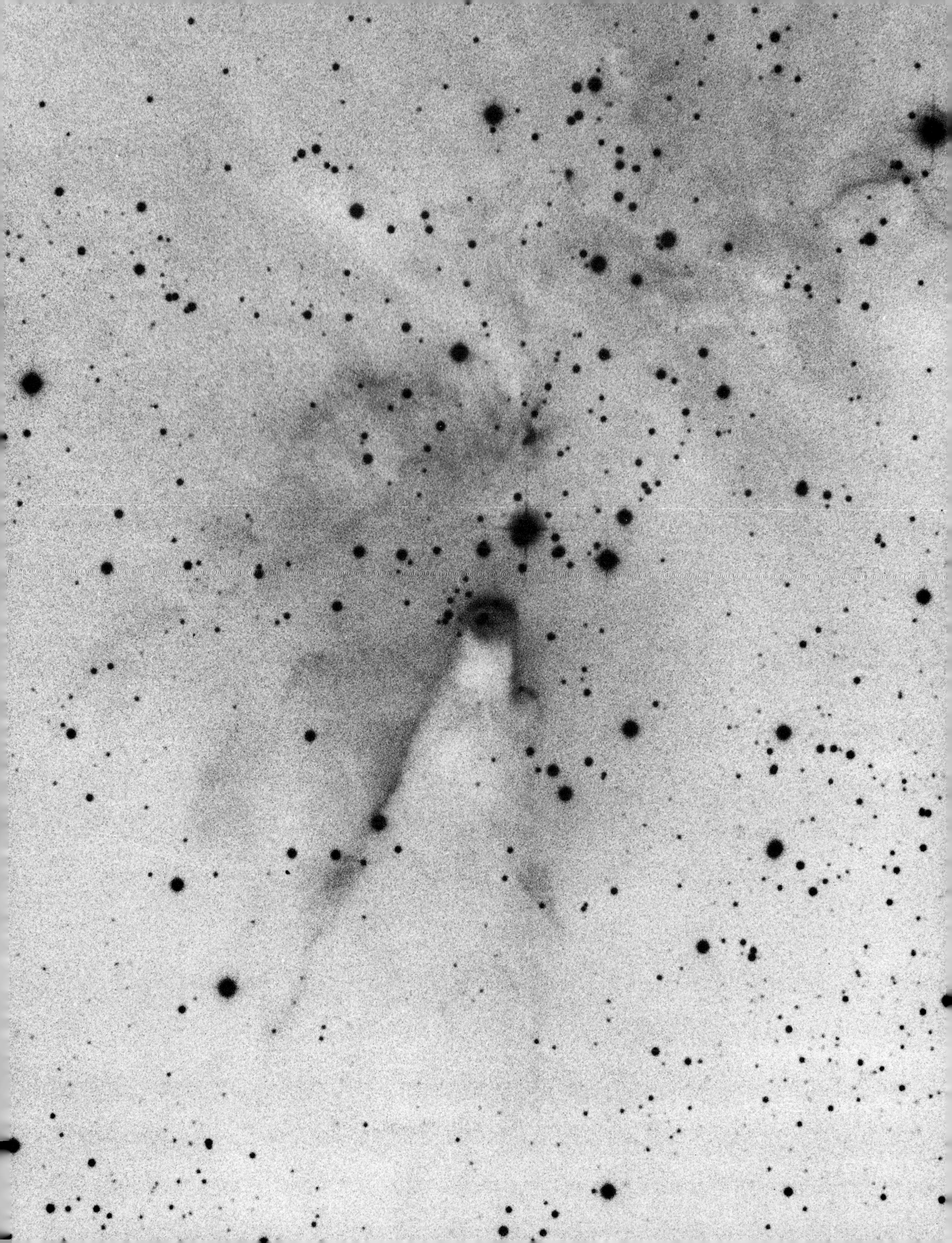

The galaxy M81

The galaxies M81 and M82 are members of a nearby cluster of galaxies in Ursa Major. M81 has a total apparent photographic magnitude of 7.8 and M82 one of 9.2. The angular distance between the two galaxies amounts to 0.6°. As the distance of the cluster of galaxies is about 3 million pc (3 Mpc), this means that their distance apart is about 30,000 pc.

A glance at photographs of M81 and M82 indicates that the types of galaxy are quite different. In the case of M81, it is easy to recognize the spiral structure. It is classed as one of the Sb-spirals, the type to which our own galaxy belongs. M81 is not an exceptional system, but, therefore, as a spiral, belongs to the largest family of extragalactic systems. M81 is of particular interest only because it is nearby, which means that it can be studied in greater detail.

M81 has attracted interest, however, in particular because of its close companion, M82. M82 is remarkable, because of its peculiar appearance. This galaxy is classified as belonging to the group of irregular, or amorphous galaxies. M82 is also one of the brightest galaxies with a total photographic magnitude of 9.2, but it is noticeably fainter than M81. M82 is an extended object with a high surface luminosity, which is divided by a large number of dark absorption lanes.

M82 became the subject of interest when, 25 years ago, long photographic exposures revealed glowing filaments spreading out from the centre, perpendicular to the main axis of symmetry of the system. These filaments are especially evident in the red light of the Hα emission line. Radial velocity measurements have revealed that the ionized gas streams away on both sides of M82 with a velocity of 1000 km/s.

In view of these observational results, it was not difficult to imagine that a powerful explosion had once taken place in M82, and that the filaments arose from the resulting ejection of material. M82 virtually came to be seen as a typical example of an exploding galaxy. Doubts were expressed about this explosion model as early as 1977. Measurements of radio emission at a wavelength of 21 cm, showed that M81 and M82 are surrounded by a common envelope of neutral hydrogen, having a total mass of 1.4×10^9 solar masses.

It can be shown that the distribution of the neutral hydrogen is determined by the effects of tidal interaction between the two galaxies M81 and M82. As a result of this interaction, gas and dust rain down on M82 and trigger exceptionally active star formation in its central region. According to this model, the centre of M82 contains immense regions of ionized hydrogen, which emit Hα radiation. This Hα radiation is scattered by the dust particles in the halo of M82, giving rise to the observed Hα filaments, The fact that the radiation of the filaments is polarized, also points to scattering processes as being the explanation.

Today, a large amount of observational material on M82 is available. Measurements have been made in the radio, millimetre and submillimetre regions, at infrared and optical wavelengths, and also in the X-ray region. From the overall spectral range the conclusion can be drawn that M82 must contain about one million hot, massive stars, which emit radiation at various frequencies. These massive stars have short lifetimes and it is thought that every five years one star in M82 will finish its life as a supernova. If the stars complete their evolution so quickly, then there must be an extremely active process of star formation, in order to constantly compensate for the rate of extinction and to maintain the radiation flux required to excite the gaseous emission.

Radio-interferometry observations with their high angular resolution, also indicate a high rate of stellar explosions. Using this observational technique, several dozen compact radio sources have been revealed in M82. The brightest compact source showed a steady decline in the total emission, which indicates that it is the remnant of a supernova that exploded recently. The observation of several dozen compact radio sources at any one time can also be explained in this manner.

It is clear from X-ray observations that the area of highest X-ray intensity is identical with the area of highest radio intensity. Strong X-ray emission can, however, also be seen on both sides of the central plane of M82. This is explained with the fact that the interstellar gas is strongly heated by the radiation from numerous hot stars, and it is expelled in directions indicated by the X-ray emission, because the stars' gravitational attraction can no longer restrain the gas. The expulsion of the gas is the result of the existence of numerous hot stars and supernova explosions.

The existence of the very high number of hot, young stars in M82 can be explained if M82 collided with M81 a long time ago. As a result, the gas density in M82 increased, leading to the rapid formation of massive stars.

On this view, the galaxies M81 and M82 are related celestial objects. For this reason, M82 will remain the focus for scientific investigation in future.

Fig. 6.22. Galaxy M81 in Ursa Major. Insert shows M81 and M82.

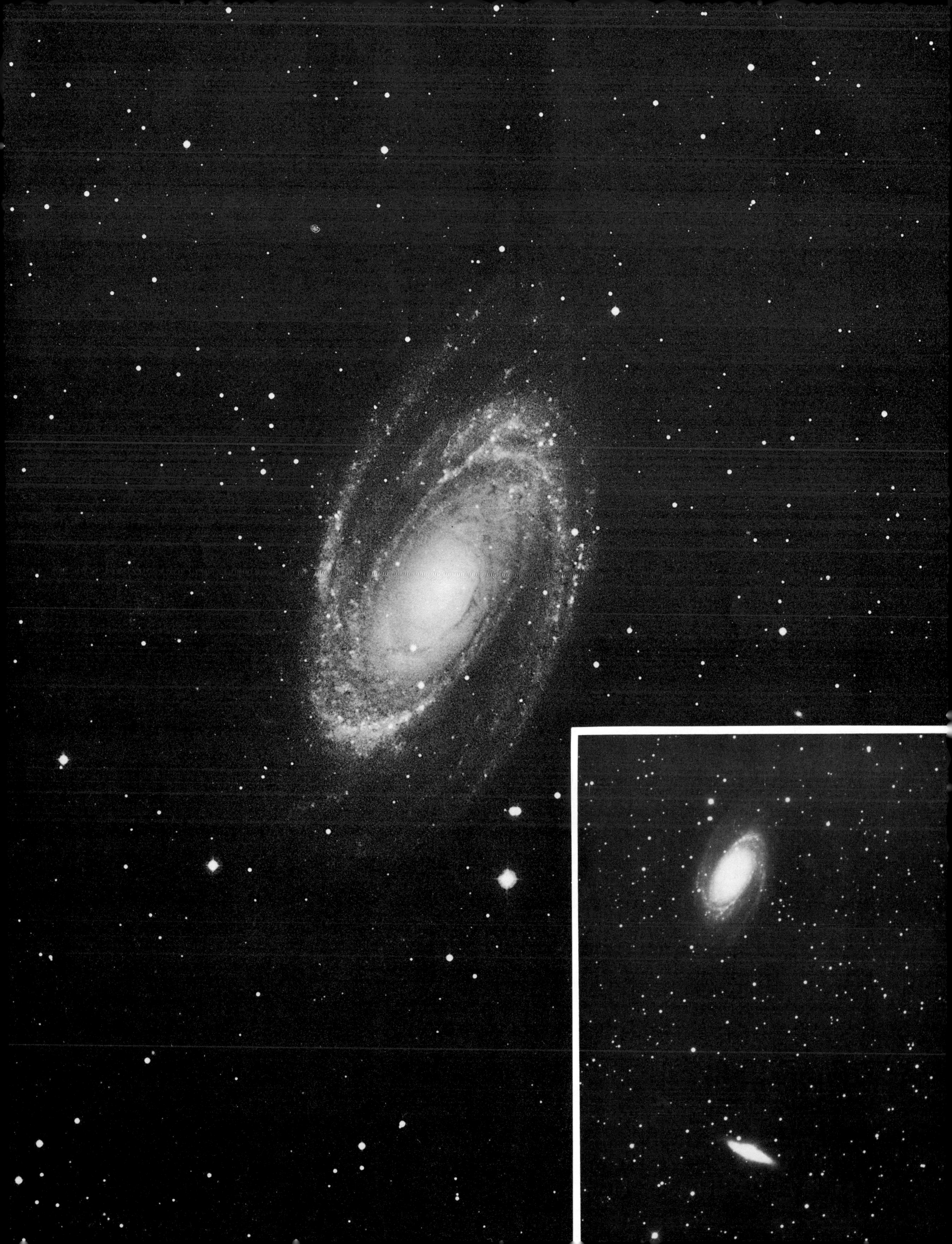

The η Carinae complex

One of the most mysterious and interesting stars in the whole sky is η Carinae in the southern hemisphere, and after which a large complex of interstellar matter has also been named. Records about this star go back as far as 1677. According to observations made by Edmond Halley, η Carinae was then of magnitude 4. John Herschel observed the star between 1834 and 1838, when he classified it as a first-magnitude object. Observations from the year 1843 describe η Carinae as having an apparent magnitude of -1, implying that it was therefore brighter than any star in the sky, apart from Sirius. Subsequently, the observed brightness of η Carinae decreased considerably. By 1869 it was invisible to the naked eye, and at its faintest, it reached magnitude 8. Since the middle of this century, the apparent magnitude has been increasing and it has now reached 6th magnitude.

Apart from the fluctuations in brightness, it was discovered in about 1920 that the object known as η Carinae is not stellar in appearance. Since then the object has continued to grow, and its dimensions are now 12×8 seconds of arc. At a distance of 2000 pc this corresponds to $25{,}000 \times 16{,}000$ AU. Some idea of the size of this object may be gained from the fact that the distance of the farthest planet from the Sun, Pluto, averages 39.4 AU. Because the elongated object resembles a short fat man with short arms and legs, it is often referred to as the 'Homunculus' Nebula. It is in a state of continuous expansion, growing at a rate of 500 km/s. How can the mysterious nature of η Carinae be explained? It is assumed that η Carinae underwent a nova-like explosion around 1840, during which the star expelled its outer envelopes. During the initial short phase, such an event can lead to an increase in brightness of several magnitudes. Dust, which is also present in the continuously expanding material that forms the Homunculus Nebula, absorbs the high-energy radiation from η Carinae. The temperature of these dust particles rises and they then radiate in the infrared spectral region. In fact, in 1970 η Carinae was shown to be the brightest object in the sky at the infrared wavelengths. This means that, inside the Homunculus Nebula, the star η Carinae itself did not change its overall radiation flux by any appreciable amount. Its optical radiation, however, is unable to reach us because it suffers extinction by the surrounding dust. These factors lead us to conclude that η Carinae and the Homunculus Nebula have a luminosity 6×10^6 times that of the Sun.

Observations also show that in the shell surrounding η Carinae there are approximately ten solar masses of dust. The latest investigations also reveal additional shell material at a distance 30 seconds of arc from the nebula, which can be explained by another explosion of the star 500 years ago.

If the star is able to eject over ten solar masses in one explosion, then its total mass must be immense. It was believed earlier that η Carinae was a star of over 100 solar masses, but the latest observations show four separate components within 0.22 seconds of arc. These may be regarded as two binaries, separated by a distance of 500–1000 AU. Three of these stars may be of 60 solar masses and one perhaps as much as 100 solar masses.

Surrounding η Carinae and the Homunculus Nebula there is a large ($2° \times 2°$) area of glowing interstellar gas which can just be seen with the naked eye. In this area there are numerous young, bright stars which excite the interstellar gas. An open cluster inside the η Carinae Nebula contains the star HD 93 129 A. This is the hottest and most luminous star known in the Galaxy. Its luminosity is more than 10^6 times that of the Sun; its mass more than 100 solar masses, and its effective temperature more than 30,000 K. In addition, a number of globules in the η Carinae Nebula are a possible indication of a future period of star formation.

The η Carinae Complex is a highly interesting area, which John Herschel fittingly described in the following way: 'It would manifestly be impossible by verbal description to give any just idea of the capricious forms and irregular gradations of light affected by the different branches and appendages of this nebula.'

Fig. 6.23. The η Carinae complex in Carina, in a negative print.

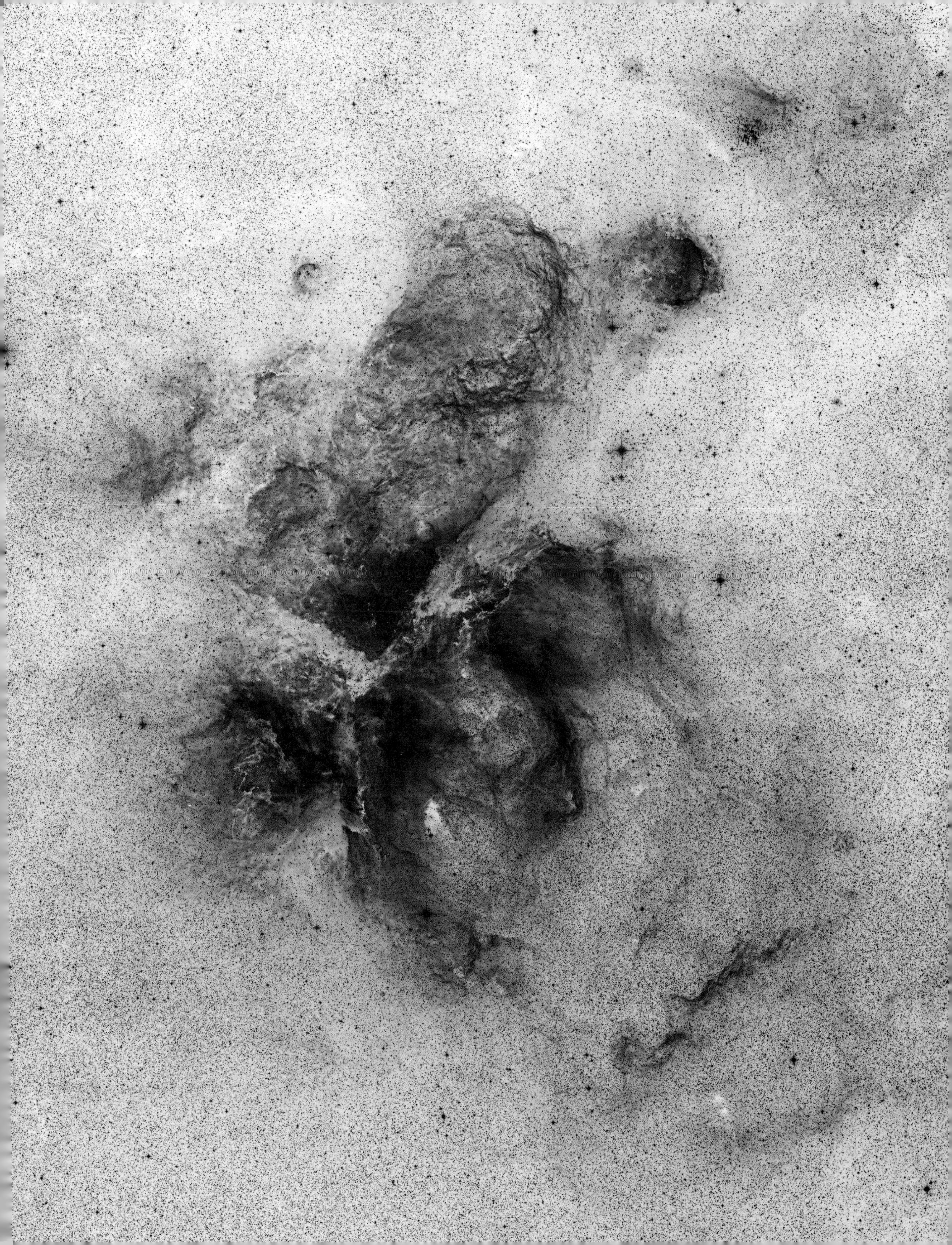

The Coma Cluster

William Herschel noticed that the nebulae that were later to be recognized as distant galaxies, were not evenly distributed over the sky. One reason for the absence of galaxies in certain areas of the sky is galactic extinction caused by the interstellar dust concentrated in the plane of our Galaxy. Even in dust-free areas at higher galactic latitudes, however, the frequency of galaxies is by no means uniform. There are clearly regions in which a higher density of galaxies occurs – i.e. galaxies tend to form clusters. Initially, we have to consider the apparent clustering, because we need to prove whether galaxies that are close to one another on the celestial sphere really are at approximately the same distance, and do form part of a related unit in space.

One way of determining the distance of remote galaxies is by using the velocity of recession of the systems. Extensive spectroscopic observations are required for this, because the shift of the spectral lines as a result of expansion of the universe as a whole has to be measured.

Many thousands of clusters of galaxies are now known. They are roughly divided into two types. The regular (or 'normal') clusters of galaxies have a central concentration, are virtually spherical and tend to be dominated by elliptical galaxies, especially at the core of the cluster. In irregular (or 'abnormal') clusters of galaxies, no central concentration is observed, and all types of galaxies are fairly evenly represented within them.

One of the best-known and most-studied, regular cluster of galaxies is the Coma Cluster, close to the North Galactic Pole. Because it is a regular cluster of galaxies, the Coma Cluster has a high density of galaxies at its core. The diameter of this central region amounts to approximately 480,000 pc. Only elliptical and S0-galaxies exist in the central region – spiral galaxies are found solely in the outer regions of the Coma Cluster. The dominant objects in the central region are the giant elliptical galaxy NGC 4889 and the giant S0-galaxy NGC 4874. In total, 1000 bright galaxies have been counted in the central region of the Coma Cluster. Several of the massive ellipticals are active galaxies. Violent activity is taking place within them that is expelling gas into intergalactic space. This intergalactic gas in the Coma Cluster appears to have a very high temperature of 100×10^6 K, because studies using artificial satellites have discovered X-ray radiation coming from the core region that corresponds to this temperature.

The Coma Cluster also contains an extended radio source. This radiation is, however, synchrotron radiation, which is caused by electrons moving at a velocity approaching the speed of light.

The velocities of recession for nearly 200 members of the Coma Cluster have been determined. The average value is 7000 km/s. This corresponds to a distance of 10^8 pc. Subsequently, the recession velocities of galaxies lying farther away from the centre of the Coma cluster were also determined, and in one particular direction, the same recession velocities were found as far as 13° away from the cluster's centre. This also applied to the neighbouring cluster, Abell 1367, which is only 17° away. The distance between the two clusters therefore amounts to nearly 3×10^7 pc. Close investigation of the radial velocities of galaxies between the Coma Cluster and Abell 1367 revealed that there is a bridge of galaxies connecting the two clusters. The groups of galaxies around NGC 3937, NGC 4065 and NGC 4213 all occur inside this bridge. The length of this bridge amounts to approximately 5×10^6 pc, indicating that a long, thin chain of galaxies stretches from the Coma Cluster to Abell 1367.

An examination of the area between 0° and 35° declination, and between 10 h and 14.5 h right ascension revealed a strong clustering of galaxies with recession velocities between 5700 and 8500 km/s. It is therefore assumed that there is a supercluster in this area at a distance of about 10^8 pc. This has been named the Coma Supercluster after the member that contains the most galaxies. The linear dimensions of the supercluster are estimated to be 40 Mpc × 40 Mpc × 100 Mpc.

Fig. 6.24. The Coma Cluster of galaxies in Coma Berenices.

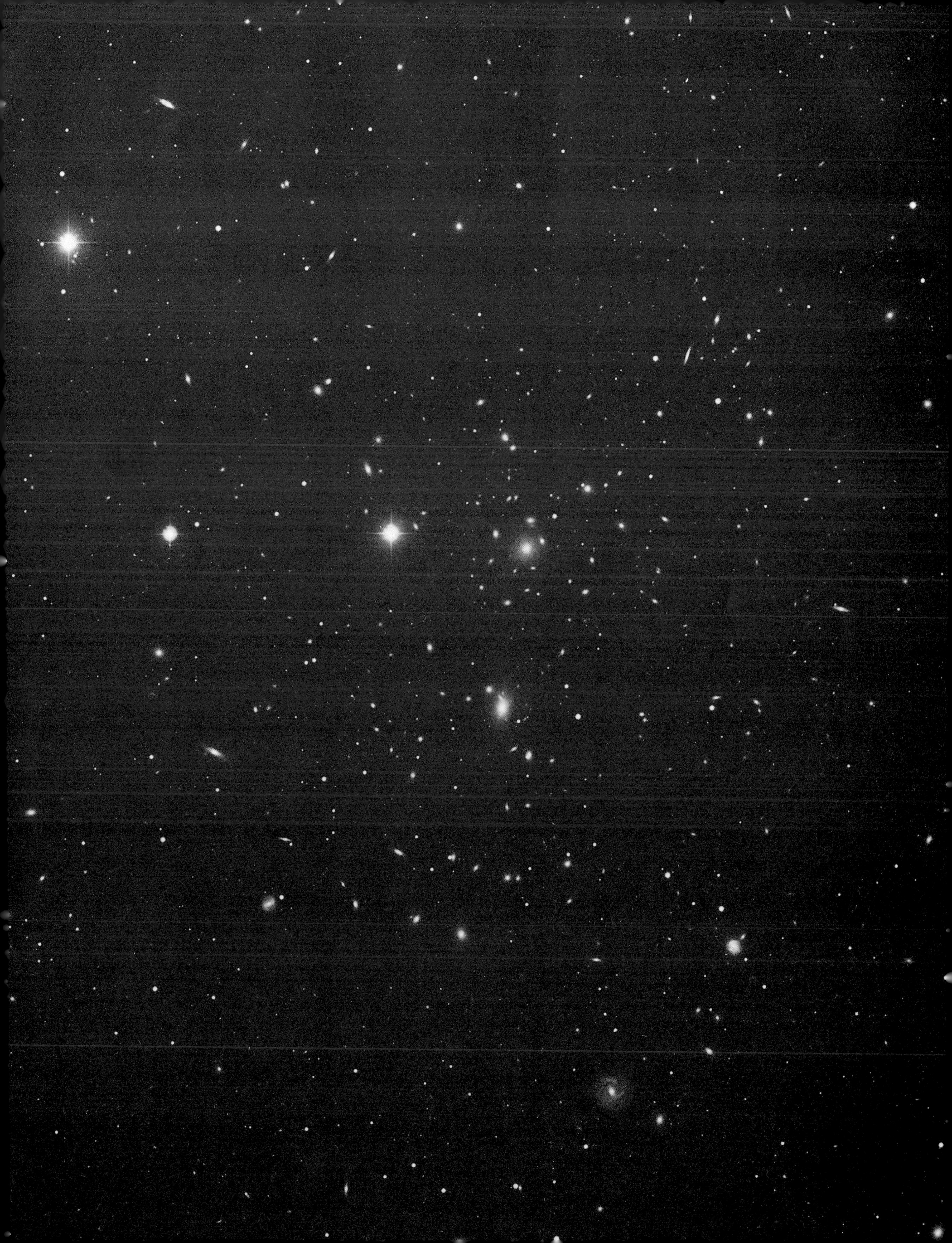

The Virgo Cluster and M87

One of the best-known and intensively investigated clusters of galaxies is the Virgo Cluster near the North Galactic Pole. Even *Popular Astronomy* by S. Newcomb and R. Engelmann says 'The most thoroughly examined object is the cluster in Coma and Virgo, which was discovered as early as 1901 by Max Wolf. It consists of several independent groups, lying at quite different distances ranging from about 2 to 14 Megaparsecs'.

The Virgo Cluster lies only 2×10^7 pc away, and its diameter amounts to about 2×10^6 pc. Its close proximity also means that it is rich in apparently bright galaxies. Sixteen of the 34 galaxies contained in the Messier catalogue are members of the Virgo Cluster. The recession velocity of the Virgo Cluster amounts to 1140 km/s. Virtually only elliptical and S0-galaxies are found in this cluster of galaxies. These are distributed almost evenly throughout the cluster and no significant central concentration is apparent. The only large-scale structure evident in the distribution is an indication of a slight, spherically symmetrical distribution of elliptical galaxies. Because the Virgo Cluster shows no pronounced central concentration, it is classified as an irregular cluster of galaxies.

A detailed examination of the Virgo Cluster and its surroundings shows that it has two extensions. One reaches south to 50° galactic latitude, and the other north to beyond the North Galactic Pole, also reaching 50° galactic latitude. At the end of the Virgo Cluster's northern extension lies the Ursa Major Cluster. Both of these clusters and the central portion of the southern extension have approximately the same recession velocity of 1100 km/s. Between 60° galactic latitude and the end of the southern extension at about 50° galactic latitude, the radial velocity increases continuously up to 2500 km/s. A correlation exists between the increasing radial velocity and the (decreasing) apparent diameter of the galaxies. Taking the apparent diameter as a criterion of distance, this means that the more distant galaxies in the extension have higher radial velocities. This suggests that the southern extension of the Virgo Cluster curves away to greater distances.

Clusters of galaxies are probably not the largest structures in the universe. From the Virgo Cluster's links with other clusters, we can draw the conclusion that there are Superclusters (clusters of clusters of galaxies), and that the Virgo Cluster is the centre of the so-called Local Supercluster. Our own Local Group, containing the Milky Way system, the Andromeda Galaxy, the Magellanic Clouds and several other galaxies also belongs to this overall grouping.

The Virgo Cluster is not only significant, however, as the centre of the Local Supercluster, but also because of one of its members, the galaxy M87, which is one of the most prominent and interesting galaxies in the sky. M87 is a giant elliptical galaxy. Its mass is thought to be between 100 and 1000 times that of our Galaxy. M87 became famous because of the large number of globular clusters in its halo. In our Galaxy's halo 120 are known; so far 4000 have been identified in the halo of M87. M87's most exciting peculiarity was discovered in 1918 by Curtis, and this is a jet of material being ejected to one side. The length of this jet is approximately 2000 pc – a particle travelling at the speed of light would need about 6500 years to go from one end of the jet to the other.

This jet was, naturally, a very important reason for the great interest that was taken in the giant galaxy, M87. The most important observational facts about the jet are as follows:

- The jet consists of individual discrete knots of material, which occur in groups. There appears to be no material between the individual groupings. The diameter of the clumps are at most 40 pc, approximately 130 light-years.
- The jet runs in a completely straight line from the core of M87 to the brightest group of knots. The line then shows waves of increasing amplitude.
- Between each individual group of knots there is a constant angular distance of 2.9 seconds of arc, corresponding to approximately 200 pc. From this we can deduce that the material in the jet does not flow continuously from the core of the galaxy, but that it is the result of periodic eruptions.
- The form and structure of the jet are the same in both the optical and radio regions.
- Spectroscopic observations have shown that the spectra of the various groups and knots within the jet are virtually identical.
- Observations in various spectral regions have shown that between optical and radio wavelengths, the radiation from the jet is synchrotron radiation. Synchrotron radiation is produced when electrons move at a very high speed in a magnetic field.
- The intensity of radiation from the jet of M87 is very high. The bright knot A alone has a luminosity 10 million times that of the Sun in the visible region.

These exceptionally interesting observational findings now have to be explained. The crucial consideration is the cause of the vast amount of energy contained in the jet. Because the jet clearly originates in the central region of M87, the galaxy itself was naturally intensively examined. It was soon discovered that M87 is a source of intensive radio emission, and as a radio source, M87 is known as Virgo A. X-ray radiation has also been discovered from M87.

Fig. 6.25. The Virgo Cluster of galaxies.

Spectral observations of M87 produced the interesting finding that, the farther away from the centre the narrower the spectral lines. The line width can however be regarded as a result of the velocity dispersion of the stars that are involved (as a group) in the production of the lines. At a distance of 72 seconds of arc from the centre, the line width showed a velocity dispersion of 230 km/s; at a distance of 9.6 seconds of arc, 278 km/s. From this variation of velocity with distance from the centre, we can deduce that a region around the centre, just 1.5 seconds of arc across – i.e. only 120 pc in diameter – contains 5×10^9 solar masses of material.

The line of ionized oxygen at 372.7 nm could be observed at the very centre. The width of this line corresponds to a velocity for the gas cloud of 1500 km/s. This line also revealed a central mass of 5×10^9 solar masses. Measurements of the surface brightness of M87 were made in parallel with these spectral observations. These indicated a mass–luminosity ratio of 6 : 1 for M87. The photometric measurements and mass-determinations for the central region, 1.5 seconds of arc across, revealed a mass–luminosity ratio of 60 : 1, however, implying that within that region equal luminosity is accompanied by ten times as much mass. This means that the extremely massive centre of M87 is very faint. One interpretation of this could be that there is a black hole at the centre of M87.

The nucleus of the galaxy may also be the source of the very fast relativistic electrons that are directed out into the jet. When these encounter the magnetic field in the knots, they would then produce the optical and other radiation. But such a conclusion would be erroneous, however, because the electrons would be greatly decelerated by the magnetic fields in the first few knots and would either not reach the distant nodes at all, or would only do so with greatly reduced velocity. Different velocities would, however, give different spectra. In fact, the observed spectra over the entire length of the jet are virtually identical.

The equivalence of spectra from all areas of the jet leads first to the conclusion that the physical conditions in all the knots must be the same, and second to the assumption that the electrons within the jet, and probably within the individual knots, are accelerated to very high speeds. Details of the acceleration mechanism remain unknown. Any investigations are made more difficult by the small size of the knots, which have a diameter of only 40 pc and also by the high energies involved.

As with M87, a single jet is also observed in the quasar 3C 273, which lies at a distance of approximately 10^9 pc.

In general, however, radio sources exhibit twin lobes. The explanation for the single jets seen in M87 and in 3C 273 is that emission from the very fast particles only occurs in the direction in which they are moving. We therefore only see the jet which is directed towards us, and not the one in the opposite direction. All in all, the Virgo Cluster has answered many questions about the large scale structure of the universe, but many still-unanswered questions have been posed by its most important member, M87.

Fig. 6.26. The galaxy M87 in the Virgo Cluster, in a negative print.

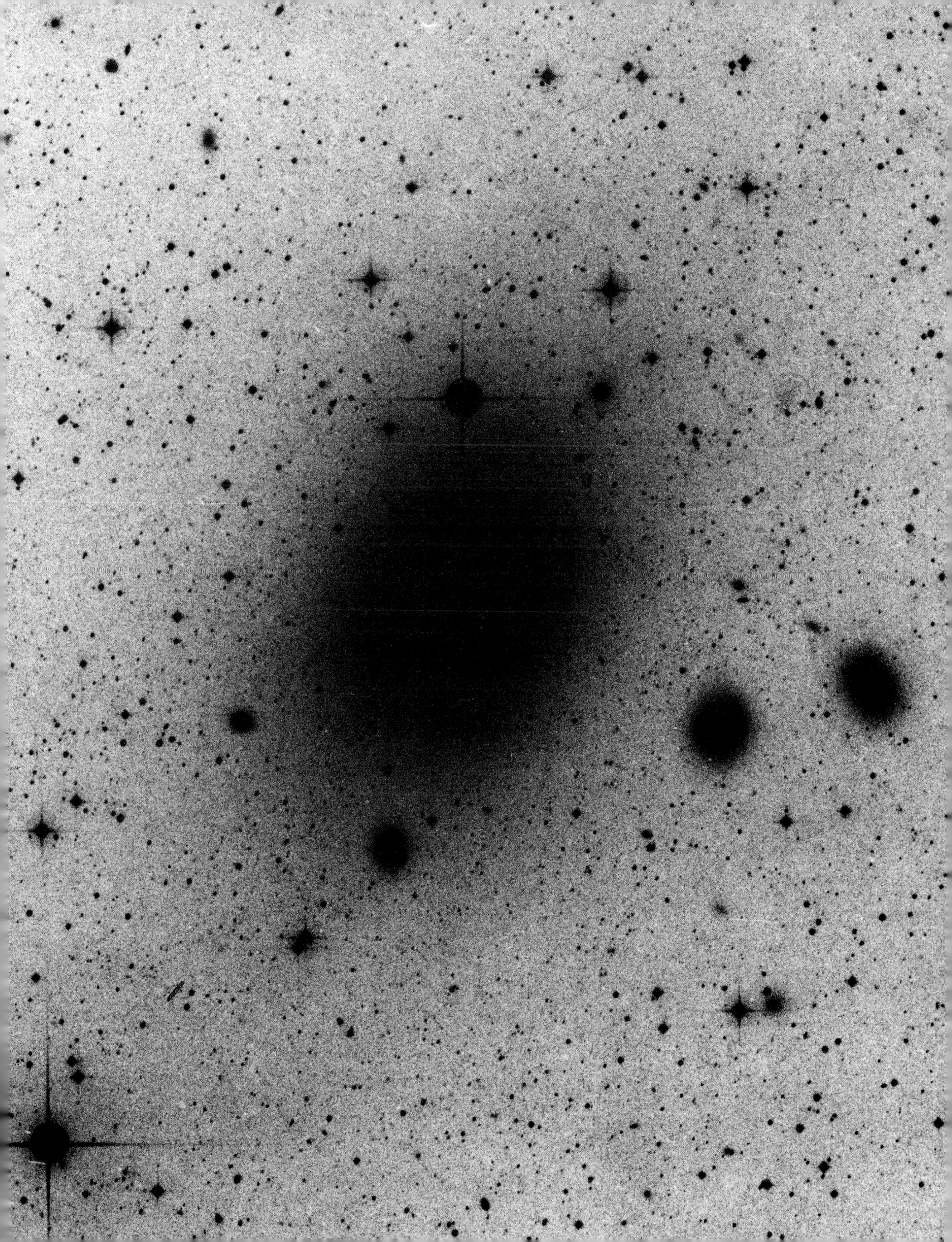

Objective prism photographs

Nearly all large Schmidt telescopes are equipped with objective prisms. In use, such a prism is located in front of the correction plate, and the telescope then obtains stellar spectra instead of direct images. As the beam of light passes from the air into the glass and then again into the air it is refracted. The extent of the refraction depends on the wavelength of the light; it is greater for short blue wavelengths than it is for the longer red wavelengths. The amount by which the light is spread out depends on the angle of refraction of the prism. The degree of angular separation is known as the dispersion, and is greater in the blue region than in the red region. In giving the dispersion for a particular prism, the wavelength at which it applies must also be stated. In calculating the final dispersion on the photographic plate, the focal length of the telescope must also be taken into consideration.

The Schmidt telescope at the Karl Schwarzschild Observatory has a focal length of 4 metres, and uses an objective prism with an angle of refraction of 0.5°. This gives a dispersion of 250 nm/mm for the Hγ Balmer line at a wavelength of 434 nm. This dispersion means that 250 nm of the spectrum correspond to one millimetre on the photographic plate. At wavelengths shorter than 434 nm, one millimetre on the photographic plate corresponds to fewer nanometres of the spectrum, and the converse holds for larger wavelengths.

It will be seen immediately that the limiting magnitude of the telescope is reduced when making spectroscopic observations with objective prisms, because the energy radiated by the stars is spread over a far greater area of the photographic plate. Taking a circular stellar image with a diameter of 50 μm – 2.5 seconds of arc in the Tautenburg Schmidt – the image takes up an area of 0.002 mm^2. With the stated dispersion of 250 nm/mm for Hγ, the length of a spectrum covering the entire spectral range is 2 mm. This means that the total area of surface exposed is 0.1 mm^2, so the spectrum appears about 50 times as large as a normal circular image. Because every object occupies a larger area on a spectroscopic plate, it is not advisable to try to photograph objects that are too faint. The reason is that the number of stars increases considerably with decreasing magnitude, and the danger exists that the spectra will overlap, and interfere with one another.

Spectroscopic observations with Schmidt telescopes are of great importance, as a few examples should indicate. For example, the special characteristic of photographic plates can be exploited. The sensitivity of plates is dependent on the wavelength. Sensitized photographic plates are not only very sensitive in the blue region but also in the longer-wavelength red region, but have a lesser sensitivity in the intermediate green region. The plate is said to suffer from 'green loss'. Stars of high effective temperature, which emit most of their energy in the short-wave blue region cause significant blackening only on one side of the green minimum; cool stars with a peak radiation at longer wavelengths darken the plate on the other side of the minimum in the green. The different densities in the low-dispersion region on one side of the minimum in the green and in the higher-dispersion region on the other side can be seen quite clearly in the various stars in this picture. Using objective prisms it is possible to distinguish easily between red and blue stars, i.e. between cool and hot stars. A black and white exposure through a prism is therefore able to give information about stellar colours.

Despite the relatively low dispersion of objective prism plates obtained at the Karl Schwarzschild Observatory, special techniques enable spectral lines to be recognized. Two or three of the strongest hydrogen absorption lines – known as the Balmer lines – can be detected. There are also celestial objects that have characteristic intensive emission lines. They emit a large percentage of their energy in narrow spectral lines. These objects can also be detected on objective-prism plates.

The objective-prism plate illustrated shows the area of the Virgo Cluster. Objective-prism plates also clearly show the difference between stars and galaxies.

Fig. 6.27. An objective-prism photograph of the Virgo Cluster.

Galaxy NGC 5128 – Centaurus A

The Galaxy NGC 5128 was mentioned by Sir John Herschel as early as 1834, when he described it as a very wonderful object. Its total magnitude of 7 means that it can also be relatively easily observed. In 1949 it was found that NGC 5128 is also a very strong radio source, and as such it is known as Centaurus A. We now know that intense X-ray radiation is also emitted from the region of NGC 5128 apart from radiation at optical and radio wavelengths.

It was not easy to determine the distance of NGC 5128 and as values proposed range between 2×10^6 and 8×10^6 pc, the mean of 5×10^6 pc is normally used. In May 1986, however, an amateur astronomer discovered a supernova in NGC 5128 a week before its maximum. Spectral observations showed that it was a Type I supernova. From the known luminosity of this type of supernova, it was easy to determine its distance, placing NGC 5128 between 2×10^6 pc and 3×10^6 pc. Therefore all other dimensions of the object, which had resulted from angular distances and the distance from our Galaxy, had to be reduced. Thus NGC 5128 is situated at the outer edge of the Local Group.

Even simple, unsophisticated photographic images of NGC 5128 indicate several peculiarities. The system, which is classified as an elliptical galaxy, is crossed by a dark band of absorbing dust that divides the Galaxy into two halves.

Following the construction of large telescopes in the Southern Hemisphere, several other interesting details were discovered. Numerous, young, blue Population-I stars were found at the edges of the dust lane and around the edge of the system as a whole. Associations of these stars similar to those in our Galaxy were also discovered in NGC 5128. Photometric observations have established that the age of the majority of the associations lies between 10 and 30 million years, but that a few are between 10^8 and 10^9 years old. By contrast, the age of the stars in the main elliptical component of NGC 5128 – which has a diameter of 20,000 pc – is about 10 billion years.

Young, hot stars are linked to interstellar matter in several ways. On the one hand, they were formed out of interstellar matter only a short time ago and remnants of the original material should still be present. On the other hand, they ionize interstellar gas clouds and cause these to become luminous. Emission from interstellar ionized hydrogen is to be expected in the Hα spectral line at 653.3 nm. Observations at this wavelength have, in fact, detected 70 luminous, ionized hydrogen regions in NGC 5128. Their distribution tallies with that of the young, hot stars, as expected. The majority of these H II regions form a ring around the elliptical component of NGC 5128. The regions of ionized hydrogen and active star formation indicate a spiral-like structure. Spectroscopic examinations of the largest H II regions show the same relative frequency of chemical elements that is found in the Orion Nebula. Emission-line observations indicate the presence of hydrogen, helium, oxygen, nitrogen, and sulphur. The enrichment of the disc of NGC 5128 in heavy elements shows that stellar evolution has occurred within it, with the resulting production of such elements, and that these have been returned to the interstellar medium during the late stages of stellar evolution.

Spectroscopic observations also show that the disc rotates around the elliptical component in a way similar to that in which spiral galaxies rotate. From this rotation, the mass inside the rotating disc has been estimated at some 4×10^{11} solar masses, which is about three times the mass of our Milky Way system. The rotation of the disc is faster than that of the elliptical component. Recent observations have shown that the elliptical component has a maximum rotation velocity of only 40 km/s.

NGC 5128 is morphologically a giant elliptical galaxy. Normally, this type of galaxy primarily consists of Population-II stars and contains little or no interstellar gas. This raises the question of why NGC 5128 has so much interstellar matter, and so many young stars. This combination of material could be explained if an elliptical galaxy collided with a spiral galaxy 1×10^9–2×10^9 years ago. It is now thought that the giant elliptical galaxy contained 3×10^{11} solar masses and that the smaller spiral galaxy involved in the collision contained only 2×10^{10} solar masses. The elliptical galaxy practically engulfed the smaller one, but a few characteristics of the spiral galaxy still remain.

This collision could also have caused the central, energetic X-ray source, and the source of the strong radio emission, that arises in the vicinity of NGC 5128. Optical observation of the central core of NGC 5128 is very difficult because of the band of dust. A jet of young stars and hot gas has however been observed streaming from the core of NGC 5128, and length of this is about 10% of the overall diameter of the elliptical galaxy. This may also be a result of the collision between the two galaxies.

Fig. 6.28. The radio source Centaurus A.

The Whirlpool Galaxy

Numerous galaxies can be seen in the constellation of Canes Venatici in the northern hemisphere. Even with simple optical aids, a typical spiral nebula can be seen, which has an apparent visual magnitude of 8.3. Because it is so easy to see, it was catalogued as M51 by Charles Messier in his catalogue of nebulae and clusters. In the 5th edition of *Popular Astronomy* by Newcomb and Engelmann (published in 1914), the nebula in Canes Venatici was described as 'the most splendid example of a spiral nebula'.

Many other famous observers examined M51 after Messier. Drawings exist by John Herschel, which he made from observations using a telescope with an aperture of 46 cm. H. C. Vogel observed this splendid spiral from Leipzig with a 20-cm refractor. His drawings, unlike those of Herschel, show the spiral structure very clearly. The most impressive drawings are those made by Lord Rosse from observations with his 183-cm telescope. These clearly show a nucleus surrounded by spiral arms. In the edition of *Popular Astronomy* mentioned above a photograph of the nebula is compared with the drawings: 'This photograph, which shows much more detail than the early drawings, tells us better than any words can the extraordinary superiority of photography over direct vision, especially in the field of nebulae. At the same time one can see just how fantastic and far from reality many of the drawings are.' The last comment particularly applies to Lord Rosse's representation of M51, which is also known as the Whirlpool Galaxy.

Photographs enable the spiral structure within M51 to be seen clearly. According to the Hubble classification the Whirlpool Galaxy is an Sc-system, i.e. the spiral arms are the dominant feature of the galaxy, whereas the nuclear region, by comparison is relatively inconspicuous. The two bright spiral arms, interspersed with interstellar dust, wind almost twice round the nucleus. In the case of M51, we see the disc of the spiral system face-on, i.e. our line of sight is directly along the axis perpendicular to the main plane of the system, and around which the galaxy rotates. M51 is about 4×10^6 pc away from us and has a radial velocity of recession of 550 km/s. The diameter of the Whirlpool Galaxy is 20,000 pc.

The most interesting thing about M51, shown by Herschel's first drawings, and by all subsequent illustrations, is the large bright knot at the end of one of the arms. The Whirlpool Galaxy itself is number 5194 in Dreyer's *New General Catalogue of Nebulae and Clusters*, and the bright knot at the end of the arm is designated NGC 5195. Nowadays, there is a generally accepted opinion about the nature of this bright knot. It is thought to be an independent galaxy, which collided with the spiral system a long time ago.

Collisions between galaxies are very rare occurrences, but they do take place. If we assume that all galaxies have masses of 10^{11} solar masses, diameters of 50,000 pc, and lie 3 Mpc from one another, then each galaxy would suffer a collision with another only once every 10^{13} years. This is about 100 times the average lifetime of a galaxy, so the vast majority of galaxies are never likely to encounter such an 'accident'. In clusters, with their higher spatial density of galaxies, the mean time between collisions is shorter.

The consequences of a collision depend on the type of collision: whether the components meet head-on, encounter one another at an angle, or just brush past. The relative velocity at the time of collision also plays an important role.

Galactic collisions can today be simulated using modern computers. Each galaxy is represented by a cluster of points representing stars and the changes in the shape of this model galaxy caused by the impact are calculated. Here the effect of close encounters between stars have to be taken into consideration, although these are rare because in actuality the distances between the stars are much greater than their diameters. But their mutual gravitational perturbations cause their positions to be altered radically.

In M51 a large, spherical, massive galaxy (NGC 5194) and a smaller, less massive galaxy (NGC 5195) were used as starting points for the computations. The spiral structure of NGC 5194, with the small galaxy NGC 5195 at the end of one arm, was the result of a grazing collision, i.e. the small galaxy passed through the outer region of the larger object.

Fig. 6.29. The Whirlpool Galaxy (M51) in Canes Venatici.

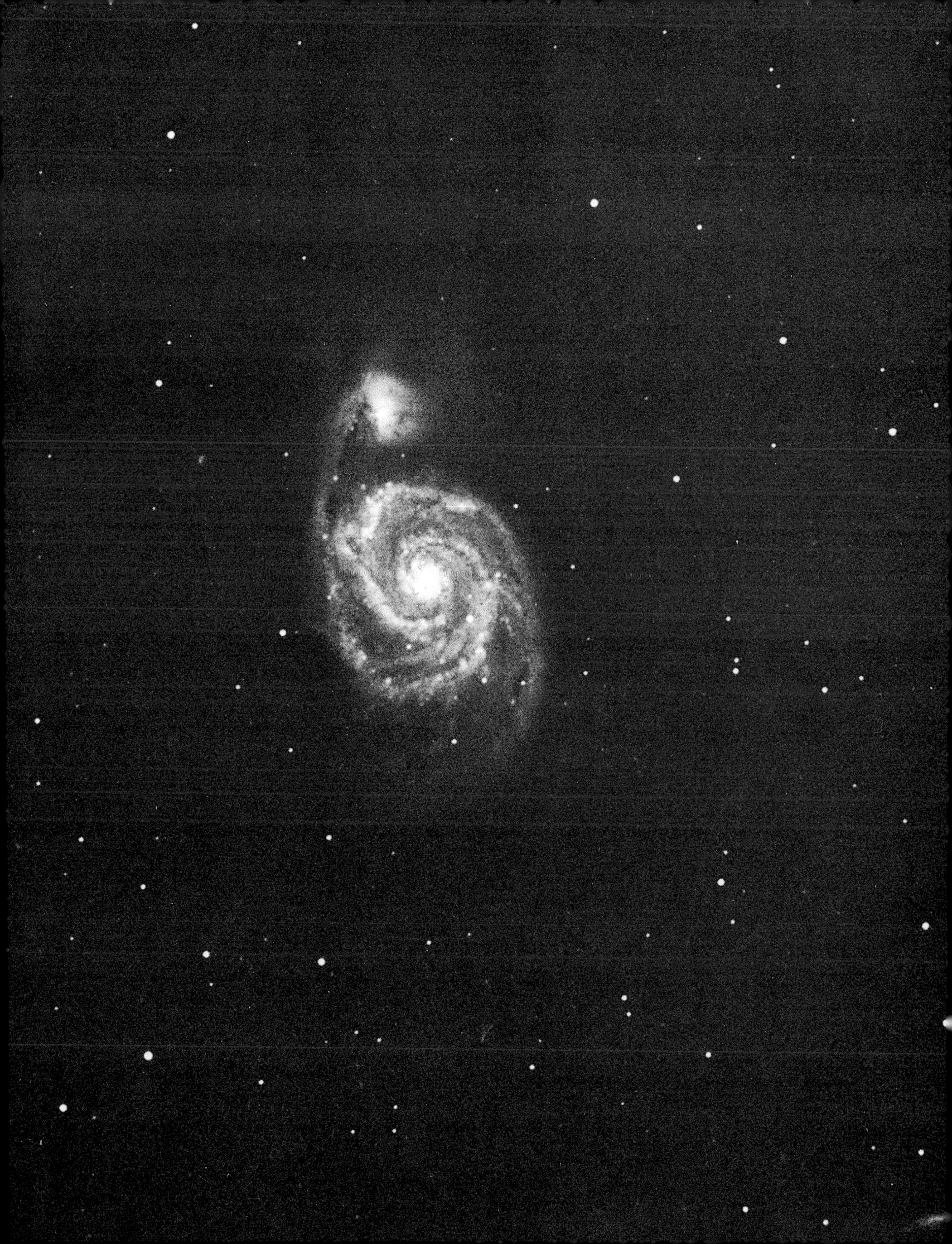

M101

The galaxy M101, NGC 5457, is a nearby stellar system. This is clear from its apparent visual magnitude of 7.9. The galaxy, which according to the Hubble classification is an Sc system, can be found on the border of Ursa Major and Boötes at RA 14 h 1.4 min and declination 54° 35′. These coordinates are for Epoch 1950.

Although the picture clearly shows that M101 is a spiral system, the galaxy is highly inclined to our line of sight. This naturally means that the apparent shape of M101 is strongly distorted.

M101 is the centre of a small group of galaxies, which contains six further extragalactic systems.

Although M101 is one of the closest galaxies, there are still uncertainties concerning its actual distance. The precise distance of M101 is, however, of great significance because M101 is an important milestone in determining the distances of more remote galaxies. Methods of determining distance are often calibrated against M101.

If its distance is incorrectly assumed, this will influence the estimated distances of other galaxies, and the discrepancy will increase with increasing distance.

The Hubble Law is a linear relationship between the distance of an extragalactic system and the redshift of its spectral lines – which is a measure of the velocity of recession. The distance/velocity ratio (the Hubble constant), is of prime importance for our knowledge of the structure and evolution of the universe. It can be determined from nearby galaxies but with only a low degree of accuracy. Their distances are reliable, but the measured shifts of the spectral lines may contain a large, individual recession component. For distant galaxies, the systematic effect of cosmic expansion predominates, but here difficulties arise in determining the true distances.

The most accurate distances of galaxies can be derived from the period–luminosity law for δ-Cephei variables. Since the absolute magnitude of δ-Cephei stars is not very high, however, this method cannot be used to determine very great distances. The most luminous δ-Cephei stars have a photographic absolute magnitude of about −5. If we assume that light from these stars suffers no additional extinction on its way to us then a star with an absolute magnitude of −5 will have an observed apparent magnitude as shown in Table 6.4 according to its actual distance.

Table 6.4. *The relationship between apparent magnitude and distance*

Apparent brightness (in magnitudes)	18	20	22	24	26
Distance (in millions of parsecs)	0.4	1.0	2.5	6.3	15.8

Table 6.4 indicates that measurements of 24th-magnitude δ-Cephei stars are required in order to determine distances out to several megaparsecs. This means that this extremely accurate method can only be used in the immediate vicinity of our Galaxy and that its particular value lies in the determination of the distance of standard galaxies, our 'milestones' in space.

In 1985/86, using modern CCD detectors at the prime focus of the 4-m telescope at Kitt Peak Observatory, it was possible to identify two δ-Cephei stars in M101, and to determine their periods. These were 37 and 47 days. It was then possible to determine the distance of M101 more accurately than before. It amounts to approximately 7×10^6 pc. This value, however, still contains a certain uncertainty. One reason for this is that the period–luminosity law is probably a period–luminosity-colour law. The influence of the colour of stars has, however, not been given much consideration as yet.

Work will therefore continue with the aim of improving our measurement of the distance of M101, in order to derive greater distances more accurately, and thus ultimately enable us to be more confident of the value of the Hubble constant.

Fig. 6.30. Galaxy M101 in Ursa Major.

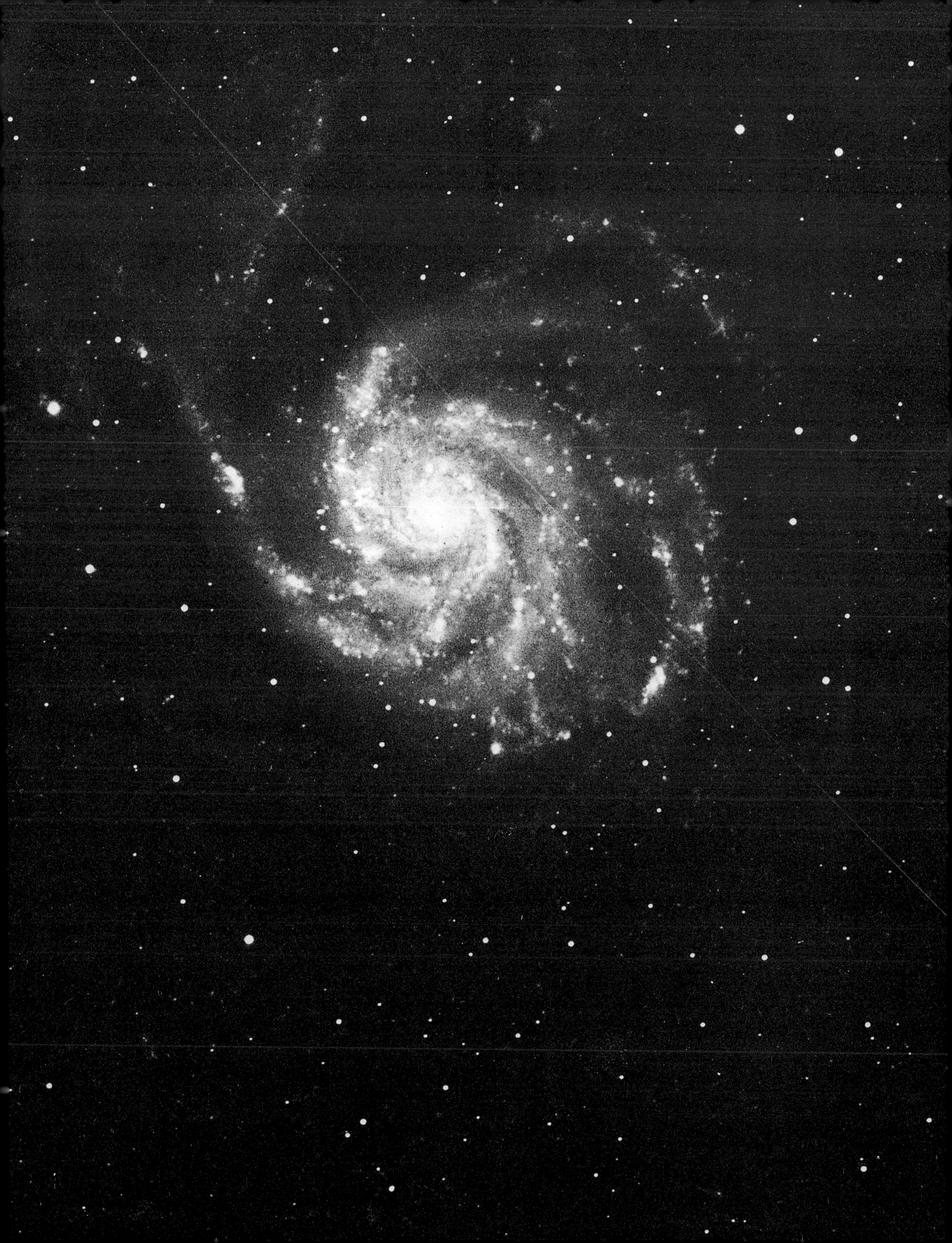

The globular cluster M92

Many of the stars in our Galaxy are single stars. One such single star is, for example, our Sun. The great majority of stars, however, 'live' in stellar communities. The smallest of these are binary systems and the largest are globular clusters. In the latter, anything between 100,000 and 10 million stars may be found. The exact determination of just how many stars there are in a globular cluster is very difficult, because individual stars can only be resolved in the outer regions of the cluster. The central areas cannot be resolved because the stars are so densely packed.

The linear diameters of the globular clusters are also difficult to deduce. In order to determine this, one requires the exact distance of the globular cluster, and an exact measurement of its angular diameter. Because the determination of the outer boundary of a globular cluster is difficult, the measurement of the angular diameter will always be relatively uncertain. In the smallest globular clusters, diameters of 15 pc have been measured and 200 pc in the largest. The majority of known globular clusters have diameters of about 30 pc. If over 100,000 stars are present in such a small volume, the density of the stars must be very high. In the vicinity of the Sun, we expect to find a density of 0.15 solar masses per cubic parsec. In other words, in a sphere with a radius of just 1.2 pc we expect to find one star of one solar mass. If we assume that in a sphere with a diameter of 30 pc there are one million stars – these figures apply to an average globular cluster – then this results in a density of 70 solar masses per cubic parsec, i.e. a density 500 times that in the vicinity of the Sun. In such a globular cluster, a sphere with a radius of only 5×10^{12} km contains one solar mass. Globular clusters have the highest stellar density of any region in the Galaxy system apart from the central core itself. If the mass of all the stars in a globular cluster were distributed over its total volume the density would be 5×10^{-21} g/cm^3. By terrestrial standards this is an unbelievably low density, but it is nevertheless very high for any region of space.

If the distribution of stars in a globular cluster is examined more closely, then it becomes apparent that the density of stars increases from the edge towards the centre. The density values given above are therefore average values for the total volume of the cluster. The density gradient from the edge to the centre can be seen in photographs of globular clusters.

Spectroscopic examinations of stars in globular clusters have shown that they are very poor in heavy elements. By 'heavy elements' we mean all elements except hydrogen and helium, the simplest elements. The heavy-element content in stars in globular clusters amounts to only 0.3%. In contrast, the content in stars in the vicinity of the Sun amounts to 2 to 3%, i.e. ten times the amount. This fundamental difference between the internal constitution of the stars in the globular clusters and those in the vicinity of the Sun has to be explained.

From the theory of the internal structure and evolution of stars, it is known that the stellar energy is produced by the fusion of lighter elements to heavier ones. Helium, for example, is produced by the fusion of four atoms of the simplest element, hydrogen. During this fusion process approximately 0.7% of the mass is lost, which according to Einstein's famous law, $E=mc^2$, is converted into energy and radiated away from the star. In the formula E is the amount of energy, m the mass and c the velocity of light. This means that, in the course of its life, a star loses a small fraction of its mass and at the same time becomes richer in heavy elements.

At certain stages of stellar evolution, stars expel part of their mass into the surrounding interstellar space. Such phases are encountered in nova and supernova explosions. In this way, heavy elements that have been formed within the star are released into the surrounding interstellar medium, from which they are incorporated into new stars at the time of their formation. The later in galactic history a star is formed, therefore, the more heavy elements it will have at the time of its formation.

Another well-known fact is that the speed of stellar evolution depends strongly on their masses. The more massive a star is, the more quickly it evolves. Massive stars therefore have a shorter life-span than less massive ones.

The stars in the globular clusters contain very few heavy elements, so they must have been formed at the very beginning of the evolution of our Galaxy, when no heavy elements had been produced by the release of energy in stars. If the members of the globular cluster were formed, however, early in the Galaxy's evolutionary lifetime, then they must have existed for a very long time, and as a result there must be a lack of very massive stars among them.

In fact, stars in globular clusters have masses ranging from 0.6 to a maximum of 1.3 solar masses. To date, the ages of 15 globular clusters have been determined by photometric methods. They were all between 15 and 18×10^9 years. Our Galaxy must be at least as old as this. Our Sun, with an age of 4.5×10^9 years, was formed during a much later stage of the Galaxy's evolution. The fact that the ages of only 15 globular clusters have been determined so far, is mainly because of their great distances. The observation of individual stars is only possible in the outer regions of globular clusters. A further problem is that low-mass stars have low absolute magnitudes. This

Fig. 6.31. The globular cluster M92 in Hercules, in a negative print.

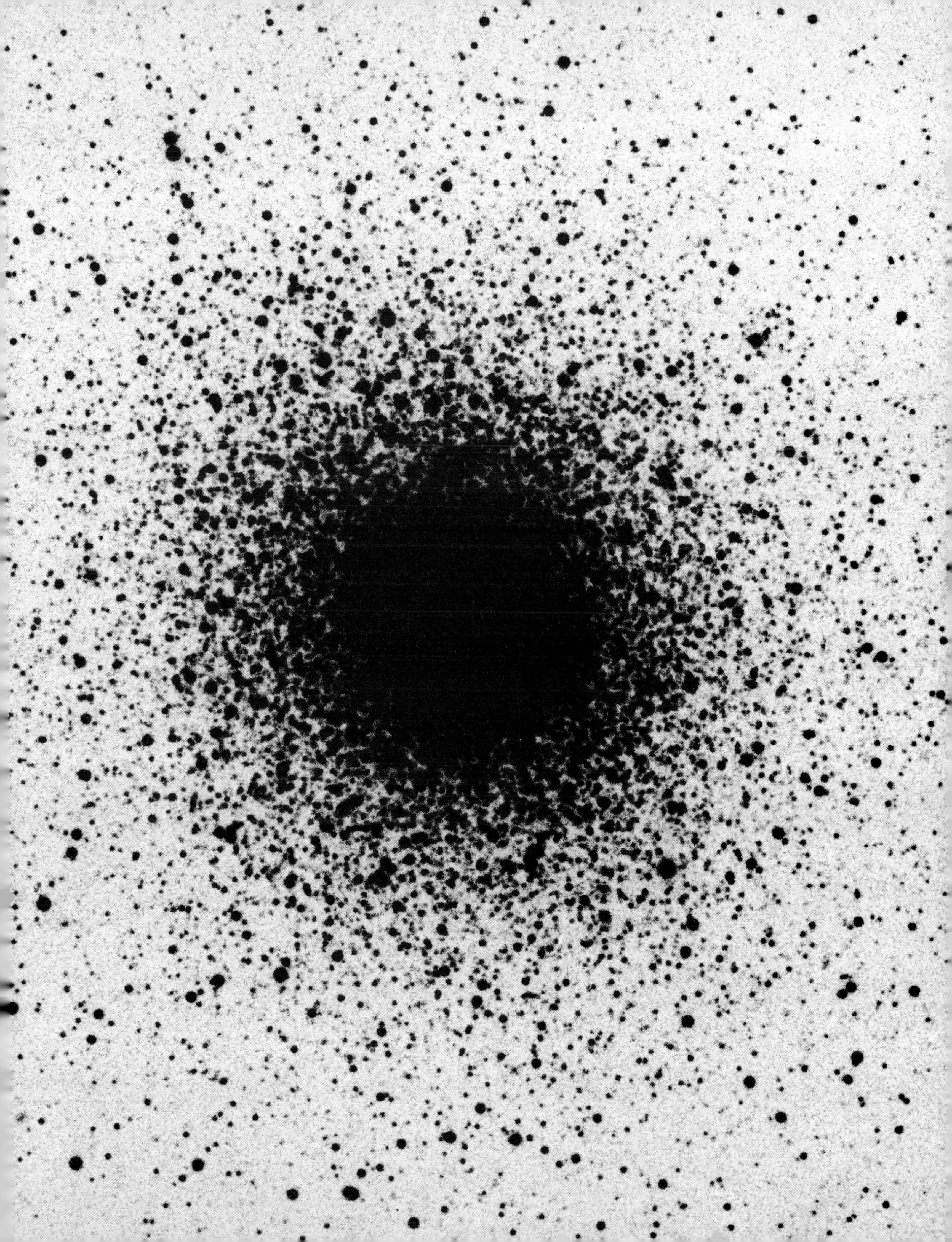

means that at their great distance they also have very low apparent magnitudes and can only be seen with the largest telescopes. The distance of the globular clusters is nearly always determined from RR-Lyrae stars. The absolute magnitude of these stars is known, and by comparing the absolute and the apparent magnitudes the distance can be calculated.

A total of about 150 globular star clusters are known to belong to the Milky Way system. Because their distances are known, it is also possible to say something about their actual distribution in space. Although the mass of the Galaxy is contained in the flat disc, the globular clusters form a large, nearly spherical shell around the flat disc. The disc has a diameter of 30,000 pc and at the centre its greatest thickness amounts to 5000 pc. In contrast, the shell formed by the globular clusters has a diameter of 50,000 pc. It is thought that there are more than the 150 known clusters, and that between 300 and 500 clusters are associated with the Galaxy. Globular clusters therfore contain between 0.2 and 0.4% of the total mass of the Galaxy.

Globular clusters have also been found in other galaxies. The Andromeda Galaxy, like our Galaxy, is surrounded by a spherical system of globular clusters.

The location of globular clusters in the Galaxy and their lack of heavy elements can be explained by the course of evolution in the Milky Way system. It is now thought that the Milky Way formed from a large spherical cloud of gas that consisted of just hydrogen and a small amount of helium. The first stars to form – including the globular clusters – still fill the spherical volume occupied by the original cloud of gas. When they were formed they only incorporated hydrogen and helium. These are the precise characteristics shown by members of globular clusters. They are the low-mass stars belonging to the first generation formed in the Galaxy. The massive stars that would have been formed at the same time finished their lives a long time ago. Because the cloud of gas was slowly rotating, it gradually collapsed, eventually forming the flat, rotating disc of the Galaxy. The objects that formed first in the Galaxy, the globular clusters, occupy a spherical volume of space, the most recently formed stars are to be found in the most strongly flattened areas of the Milky Way system, in the disc's plane of symmetry.

Investigation of globular clusters therefore tells us a lot about the early evolution of our Galaxy, and also of other galaxies, because globular clusters are the oldest members of stellar systems.

Fig. 6.32. The globular cluster M92 in Hercules. A negative print after special photographic treatment.

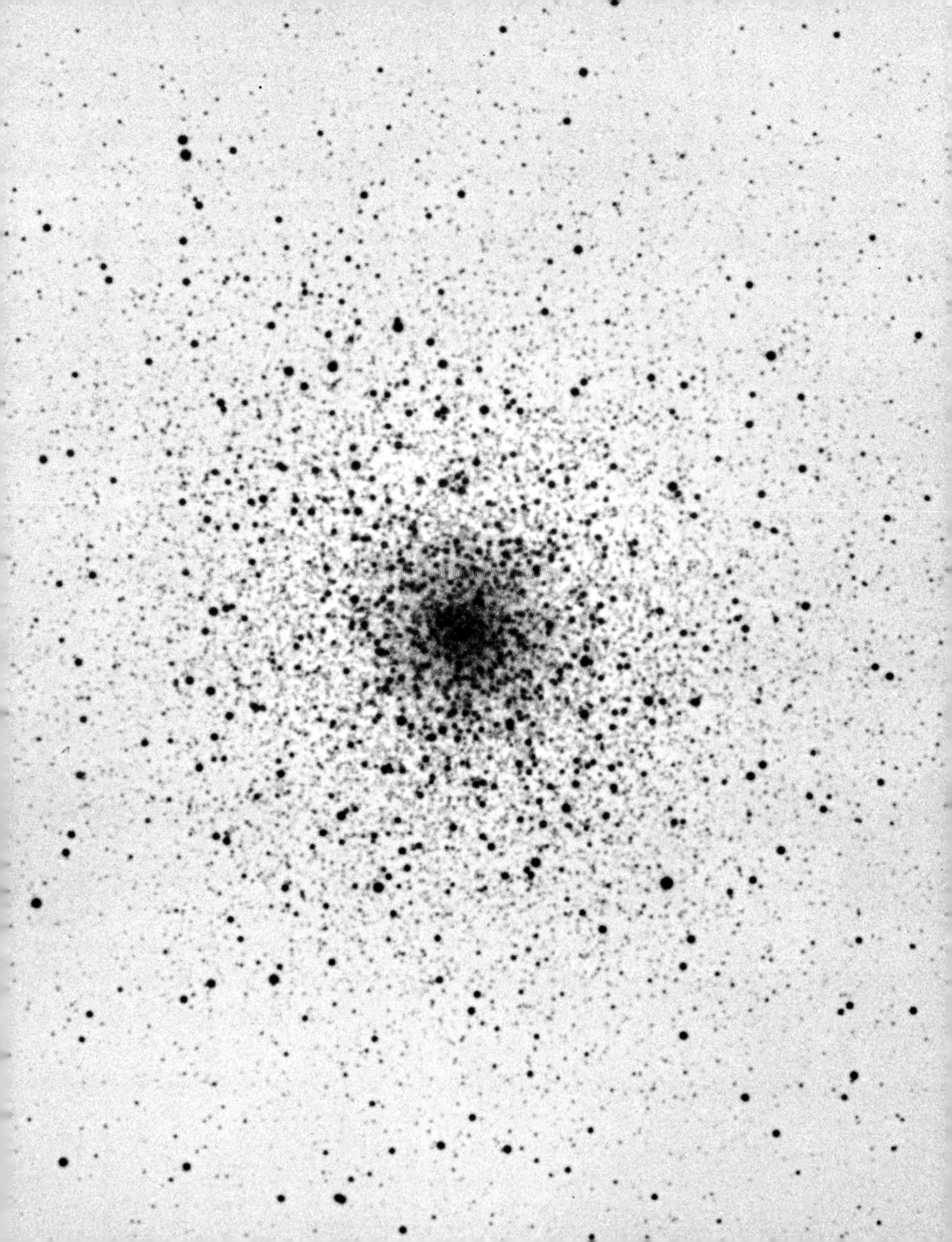

The Lagoon Nebula

Many of the celestial objects that arise both from the effects of the distribution of interstellar gas and dust, and from their interaction with various sources of radiation, are frequently very impressive. Only a few of these luminous 'nebulae' can be seen with small instruments, binoculars or the naked eye. One of these rare objects is the Lagoon Nebula (NGC 6523, M8) in Sagittarius. This bright nebula, which is conspicuous in the photograph, lies in an area of the Milky Way that is rich in stars. There are further examples of luminous interstellar material in its vicinity. To the north is the Trifid Nebula (NGC 6514, M20) with its characteristic division into three, and at the eastern edge of the picture, parts of the complex that includes NGC 6559, IC 1274 and IC 1275. The name 'Lagoon Nebula' is derived from a band of absorbing material that appears in projection against the luminous background. This characteristic feature is hardly visible visually, however, and only appears clearly on appropriately exposed photographs. In the illustration shown here it has been overwhelmed by other, brighter parts of the object. The lagoon can be vaguely recognized as a darker band running from north-east to south-west across the main area of luminosity.

An important source of the nebula's luminosity is the young cluster NGC 6530, visible as a small concentration of brighter stars on the north-eastern edge of the lagoon. Astrometric methods show that the members of the cluster occur over the entire area of the nebula, even including the fainter, outer zones. The distance of the cluster has been reliably estimated at 1800 pc. Because of the relationship between the cluster and the bright gas, this value simultaneously gives the distance of the Lagoon Nebula, and also indicates that it lies in the Galaxy's Sagittarius arm close to the Sun.

The brightest parts of the Lagoon Nebula are excited by stars embedded within it that have a particularly high surface temperature of between 30,000 and 40,000 K. One of these stars is Herschel 36. It lies to the west of the dark band but is quite inconspicuous because of the absorbing material. Herschel 36 is remarkable because it is the energy source for the Hourglass Nebula. This is a small elongated nebula, narrower in the middle, which is the first sign of the Lagoon Nebula on short-exposure photographs. Sir John Herschel observed it but, in ignorance of its true nature, had regarded it as a spiral nebula.

The Lagoon Nebula is extremely young. Based on calculations concerning stellar evolution, the associated cluster NGC 6530 may be given an age of five million years at most. Measurements carried out in the infrared region and at radio frequencies also clearly indicate the presence of stars in the formative stage. Young stars and luminous interstellar material in the form of gas or dust always appear together, and both are generally associated with extended complexes of very dense, cool gas. In the case of the Lagoon Nebula the molecular cloud – the physical properties in the dense gas promote the formation of molecules – lies farther along our line of sight, behind the optically luminous nebula.

Because of its great surface brightness, the Lagoon Nebula offers an opportunity of carrying out a high-resolution investigation of the internal motion of the gas. Because emission from the gas takes place almost exclusively in sharp spectral lines, the radial components of the motion can be determined spectroscopically from the Doppler effect. The result is a complicated picture. The apparent structure of the Lagoon Nebula, concentric to the cluster NGC 6530, suggests a bubble expanding away from the centre. Investigations fail to support such a simple idea. The shape of the Hourglass Nebula suggested that the explanation might be a bipolar flow of gas outwards in opposite directions, and some confirmation for this idea came from findings from other young objects. But, once again, the data fail to support this theory, and the bipolar structure of the Hourglass Nebula is possibly merely related to the distribution of absorbing material.

Overall, the velocity measurements indicate a systematic motion of the nebular gas away from the molecular cloud at a speed of about 10 km/s. This can also be understood in physical terms. Statistics show that the majority of regions of ionized gas are observed at the edges of molecular clouds. The cause of these H II regions are newly formed, hot stars. The ionization of the surrounding gas by the stars' short-wave radiation creates a shock wave, which forces the surrounding gas to expand. In regions of high density gradients, as on the edge of molecular clouds, this expansion takes on the character of a supersonic flow at a high Mach number, and ionized gas erupts from the molecular cloud like a geyser. This state is appropriately known as the 'Champagne phase' in the evolution of an H II region. The pattern of the flow in the Lagoon Nebula seems to be of just this type.

Fig. 6.33. The Lagoon Nebula in Sagittarius.

The Omega Nebula

This cloud of ionized interstellar gas is one of the brightest objects of its kind, and has been the subject of particularly detailed study in recent years. The reason for this is that the region has recently been the site of star formation. The non-stellar material is also very favourably placed with respect to the line of sight for this process to be examined and, in addition, this area of the sky is accessible to all the large optical, radio and infrared instruments.

The Omega Nebula offers an impressive example of the increase in knowledge that takes place when we are able to extend the range of wavelengths. Each detector, which is only effective over part of the spectrum 'sees' a different picture. These may be combined to give an idea of the whole complex and of the way in which the various components are distributed in space, and also the changes in stellar formation and evolution with time.

To the west, the Omega Nebula borders on a dense molecular cloud. Information about its physical properties can be determined by radio observations of molecules such as carbon monoxide, formaldehyde, and ammonia. Especially interesting is the densest part of this cloud, which lies to the south-west of the nebula. Here, in a region having a diameter of 3.5 pc, there is a concentration of about 10,000 solar masses of material, predominantly H_2 gas. Compact infrared sources and maser emission are an indication of current star formation. Optically, and in the picture, this material is invisible. The stars that are visible in this direction are foreground objects, and no visible radiation is able to penetrate the cloud from anything in the background.

A further interesting component can only be partially identified at optical wavelengths. This is a region of hot, O-type stars, lying at the edge of the molecular cloud that is closest to the visible nebula. Compared with the molecular cloud, these stars form a slightly older generation of stars, and they are responsible for both the intense radio emission in the centimetre and decimetre regions, and the radiation at optical wavelengths, which give rise to the Omega Nebula itself. The radiation originates in the ionized-hydrogen zone, the H II region, which the hot stars on the edge of the molecular cloud create around them.

Radio observations reveal two elongated maxima measuring 6 pc by 1.5 pc, one of which lies tangentially to the south-western edge of the nebula, and coinciding with the position of the hot stars. The optical quanta that arise in the H II region are completely absorbed by the dense material in the molecular cloud. Stars are also subject to this extinction effect and may show a decrease in brightness of more than eight magnitudes. The other maximum of radio emission coincides with the bright, southern part of the nebula. Here, because of lower extinction, the optical radiation is also visible.

The density of electrons in the H II region amounts to 10^3 cm^{-3} but there are notable internal concentrations with diameters of less than 0.3 pc and densities up to 10^5 cm^{-3}. These are small pockets of neutral hydrogen, which are only ionized on their surface by stellar radiation. They have survived as remnants of the molecular cloud that has been destroyed by the formation and evolution of hot stars. Apart from constant 'erosion' of their surfaces, these objects appear to be stable and have no tendency to evolve into stars. However, the tangential region revealed by radio observations contains a compact source, 0.4 seconds of arc across, which emits radio and infrared radiation. This object is thought to be a recently formed star, of spectral class B0, where the infrared emission comes from dust heated by the stellar radiation.

The molecules in the molecular cloud and the gaseous ions in the H II region emit radiation in the form of spectral lines, and can be used to study any motion that is present by means of the Doppler effect. The results of such investigations suggest that a stream of ionized gas is flowing out of the molecular cloud at right angles to the line of sight and falling on the Omega Nebula. The whole complex around M17 is therefore an example of where stars have formed at the boundaries of a molecular cloud, and which are now visible as hot O- and B-type stars. They produce an extended H II region around them, which, because of its higher gas pressure, tends to expand in the direction where the surrounding medium is less dense. In addition, the shockwave that propagates back into the molecular cloud promotes the formation of a new generation of stars.

The Omega Nebula lies at a distance of 2200 pc, and belongs to the spiral arm closest to the Sun in the direction of the galactic centre, the Sagittarius arm.

Fig. 6.34. The Omega Nebula in Sagittarius.

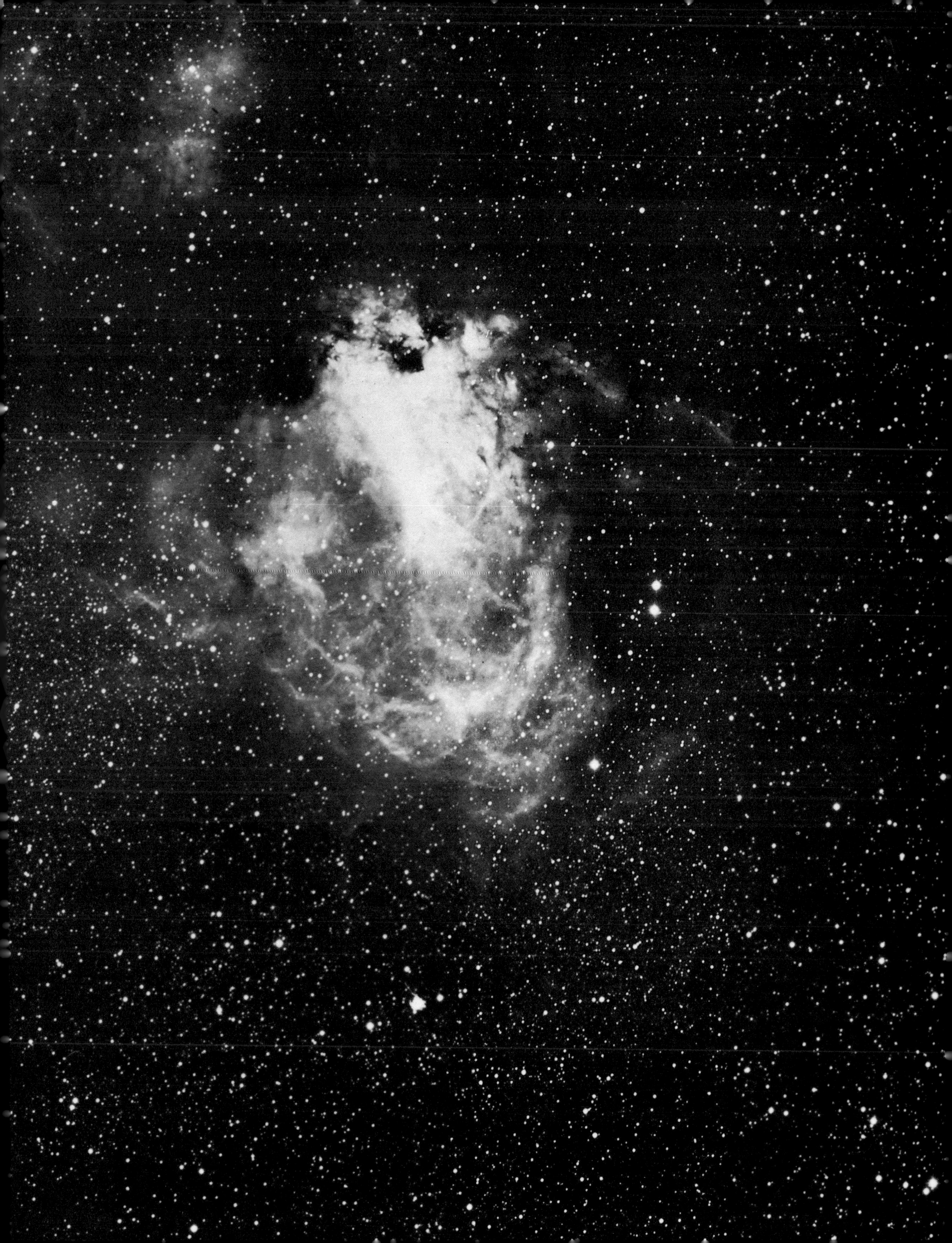

The Dumbbell Nebula

The Dumbbell Nebula is one of the planetary nebulae and, in fact, it was the first one known. In 1764 Charles Messier included it in his list of unusual celestial objects. The term 'planetary nebulae', used for this type of bright, gaseous object, refers to the resemblance to a planetary disc that is frequently found when these objects are viewed through a telescope. Their scientific study began in 1918, when a whole volume of the publications of the Lick Observatory on Mt. Hamilton in California was devoted to 'studies of the nebulae' undertaken with the 90-cm Crossley Reflector. This contained a paper by H. D. Curtis, based on photographs of 78 planetary nebulae, which described them in detail, classifying them according to their shape, their distribution on the sky, and the inner kinematics. It also included consideration on a classification by evolutionary state, although such ideas have now been superseded. From the large number of circular, or at most, only slightly elliptical, discs or rings, the justifiable conclusion was drawn that planetary nebulae must be spherical or ellipsodial shells. In the work mentioned, the Dumbbell Nebula is classified among the discs that show brighter edges and incomplete structure. Because its angular diameter is 8 minutes of arc, it counts as a giant among planetary nebulae and, because of its clearly visible structure, it is particularly suited for theoretical investigation of the phenomenon.

One characteristic of a planetary nebula is the occurrence of a central star, which tends to emphasize the symmetrical appearance and which is also the actual centre of the object. The central stars are remarkable because of their particularly high surface temperatures ranging from 40,000 K to more than 100,000 K. Although their individual spectra are often very different, they generally show emission lines of hydrogen, helium, carbon and nitrogen. A planetary nebula is an expanding shell of gas around its central star. The expansion velocities, measurable by the Doppler effect on the spectral lines, lie between 10 and 40 km/s. The slower shells also have smaller radii of less than 0.1 pc. The gases are ionized by the intense ultraviolet radiation from the hot stellar surface. This explains the occurrence of emission lines in the spectra. The density of ions in the shell is of the order of 100,000 particles/cm^3.

In 1956, the Soviet astrophysicist, Iosif Shklovskii, proposed the hypothesis – which has since been generally accepted and confirmed – that planetary nebulae are the shells that have been ejected from stars in their red-giant phase. This advanced evolutionary stage is characterized by a specific distribution of chemical elements and of energy-production zones within the star. It is also evident externally, among other things, by a continuous ejection of mass. This mass-loss, known as a stellar wind, amounts to at least 10^{-6} solar masses per year, although a 'superwind' could be over 100 times this amount for a short time. The physical causes of the stellar wind are still not completely understood. Model calculations of evolved stars show, however, that instabilities occur in their outer layers, which may at least have the effect of providing the initial impetus.

As a result of this mass-loss, which occurs throughout the giant phase, stars evolving towards the planetary nebula stage fairly consistently shrink from an original mass of between one and four solar masses to about half a solar mass. This is a typical value for the central stars. In this process, the star has gradually lost a large proportion of the material in its outer layers, which are rich in hydrogen and not affected by nuclear processes. The layers that remain, which once lay deep in the interior, now influence the star's spectrum. In particular, the effective temperature of the stellar surface rises rapidly to the very high values mentioned above. This causes the surrounding gas, which, because of its low velocity, is still close to the star, to be ionized. When ionization occurs, there is a strong increase in pressure, and a corresponding acceleration of the rate of expansion of the inner edge of the shell. This, in turn, tends to sweep up additional gas, thus forming a relatively dense shell and giving rise to a planetary nebula.

The evolutionary stages that occurred before the nebular phase consume so much energy from the star's resources that the remnant left as the central star does not have a very long life time. Within a few tens of thousands of years its temperature will drop so much that the supply of ultraviolet photons required to maintain ionization of the gas will fail to be produced. The existence of the planetary nebula is therefore limited to a period of a similar order of magnitude. It was once thought that these aesthetically beautiful planetary nebulae were rare, but we now know that they represent a normal stage in the evolution of many stars, albeit one that is not very often observed because of its short duration.

Fig. 6.35. The Dumbbell Nebula (M27) in Vulpecula.

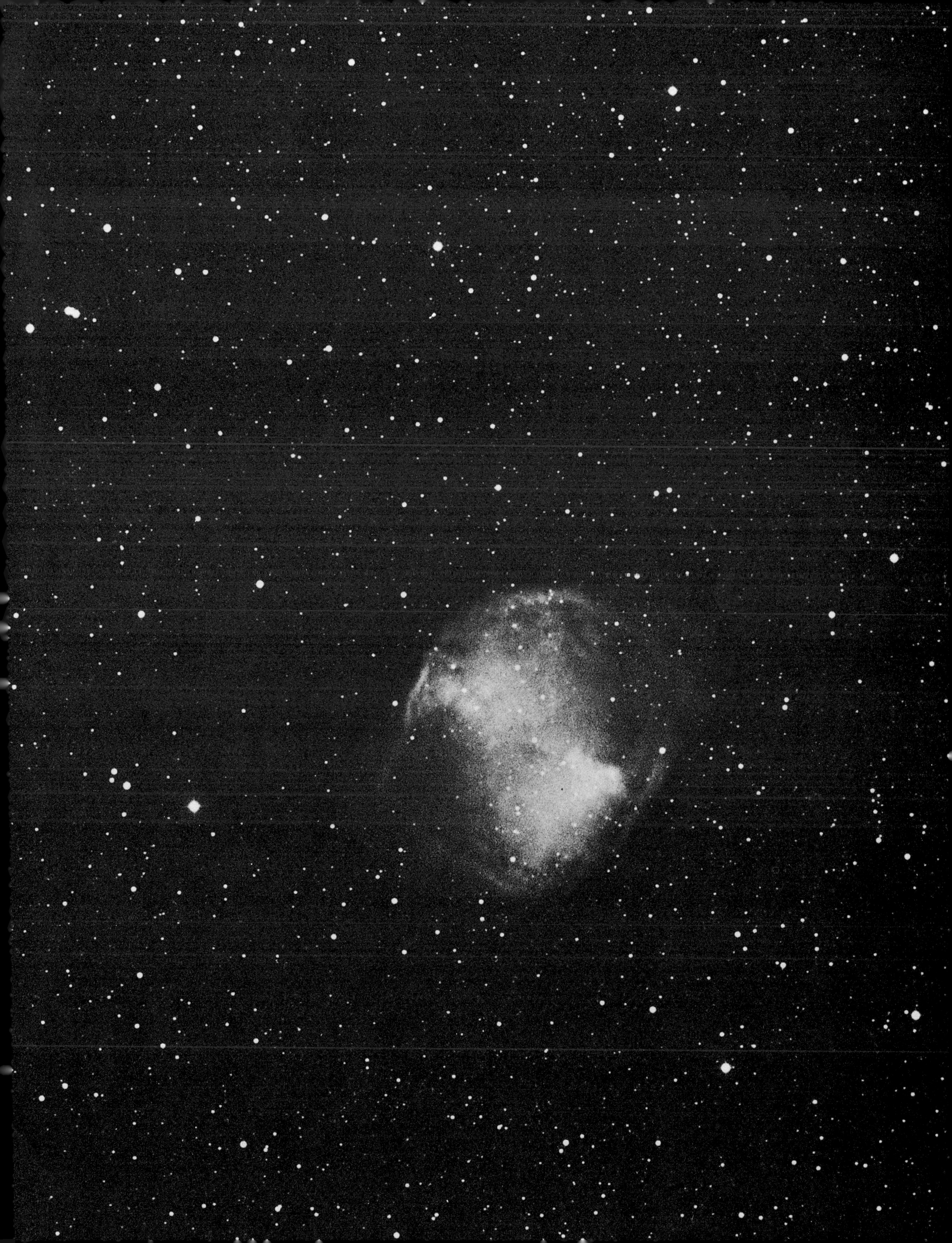

The area around the star γ Cygni

The area around the bright star γ Cygni represents an ideal example of the way in which various forms of interstellar material frequently accompany one another. We find that the ragged structure in the bright nebulosity, which is itself caused either by emission from gas or by scattering from widely dispersed solid particles, is closely linked with dark areas that have equally irregular boundaries. The surface brightness in the luminous areas ranges from being just above the limit of detectability, to exceptionally bright, sharp boundaries. We can also see globules – sharply defined concentrations of light-absorbing material with diameters of only a few seconds of arc – projected against the bright regions.

The discovery of bright galactic nebulae began with the introduction of telescopic observations. In 1610, the Frenchman, Nicholas Peiresc, was the first to mention the Orion Nebula. There are a few bright gaseous nebulae that can be seen with the naked eye, but it is necessary to know their position beforehand. Without optical assistance it would be extremely difficult to discover such objects. Irregularly shaped, nebula-like objects in the sky were of course well known. With the use of the telescope these could generally be resolved into individual stars and recognized as clusters. This was the basis for the assertion, which was made until well into the nineteenth century, that with adequate telescopic resolution all nebula-like objects in the sky would be shown to be tight clusters of stars. Even the experienced William Herschel first supported this opinion, and explained the formation of 'nebulae' as actual concentrations of stars, which, from being originally scattered in space, had clustered together as a result of their own gravitation. The observation of a planetary nebula – as we would now call it – with a central star and luminous shell, convinced Herschel that nebular matter actually existed, both associated with stars and also independent of them. Such realistic ideas still did not signify an end to the long-running debate on the possibility of resolving nebulae into individual stars. The existence of true gaseous nebulae was first established when, in 1864, William Huggins examined various objects using visual spectroscopy, and recognized pure line spectra of luminous gas in some of them. In particular, he discovered the Hβ and Hγ of the Balmer series of hydrogen, and the conspicuous N_1 and N_2 lines, which could not be interpreted, and were therefore attributed to a hypothetical element called 'Nebulium'. The puzzle was not solved until 1928 by Ira S. Bowen, who recognized the mysterious lines as emission from doubly ionized oxygen atoms in a 'forbidden' state that cannot be attained in a terrestrial laboratory.

The use of photography in astronomy led to a great increase in the knowledge of the existence and the variety of shapes shown by these objects, and also about their physical nature. Pioneers of nebular photography who are worth particular mention, are E. E. Barnard, who first worked at the Lick and later at the Yerkes Observatory, and Max Wolf of Heidelberg. Both discovered and described numerous such objects. At first the pictures were taken with lenses, that had apertures of 15 centimetres or even less. Their suitability for nebular photography arose from their large fields of view.

Although spectroscopy provided early proof of the gaseous nature of many bright nebulae, the often closely related 'stellar voids' were the subject of discussion for a long time. Wolf contributed to the idea that material that absorbed light was situated in front of an otherwise unaffected field of stars. By carrying out star counts in and around the dark clouds, he showed that, beyond a certain value, the number of stars in the dark areas generally increased more slowly towards fainter magnitudes than in an unaffected field. He attributed this to the screening effect of layers of absorbing material lying in the foreground. We now know that this material consists of sub-micrometre-sized particles with refractory nuclei, often coated with volatile, molecular compounds. This dust-like material is not only noticeable because of its extinction of light, but also because (under appropriate conditions) its scattering effects can lead to the formation of reflection nebulae. Here again, it was early nebular spectroscopy that led to the discovery of the cause of the luminosity of this type of nebula, when the spectra of the nebular material was found to correspond with that of neighbouring stars.

Interstellar matter, such as that which appears so impressive around γ Cygni, is still the subject of intensive research. Questions concerning the detailed chemical composition, the physical state, the energy balance, the interaction with other galactic components, have proved to be of prime importance as have more general issues about the formation and evolution of this type of cosmic material.

Fig. 6.36. The area around the star γ Cygni.

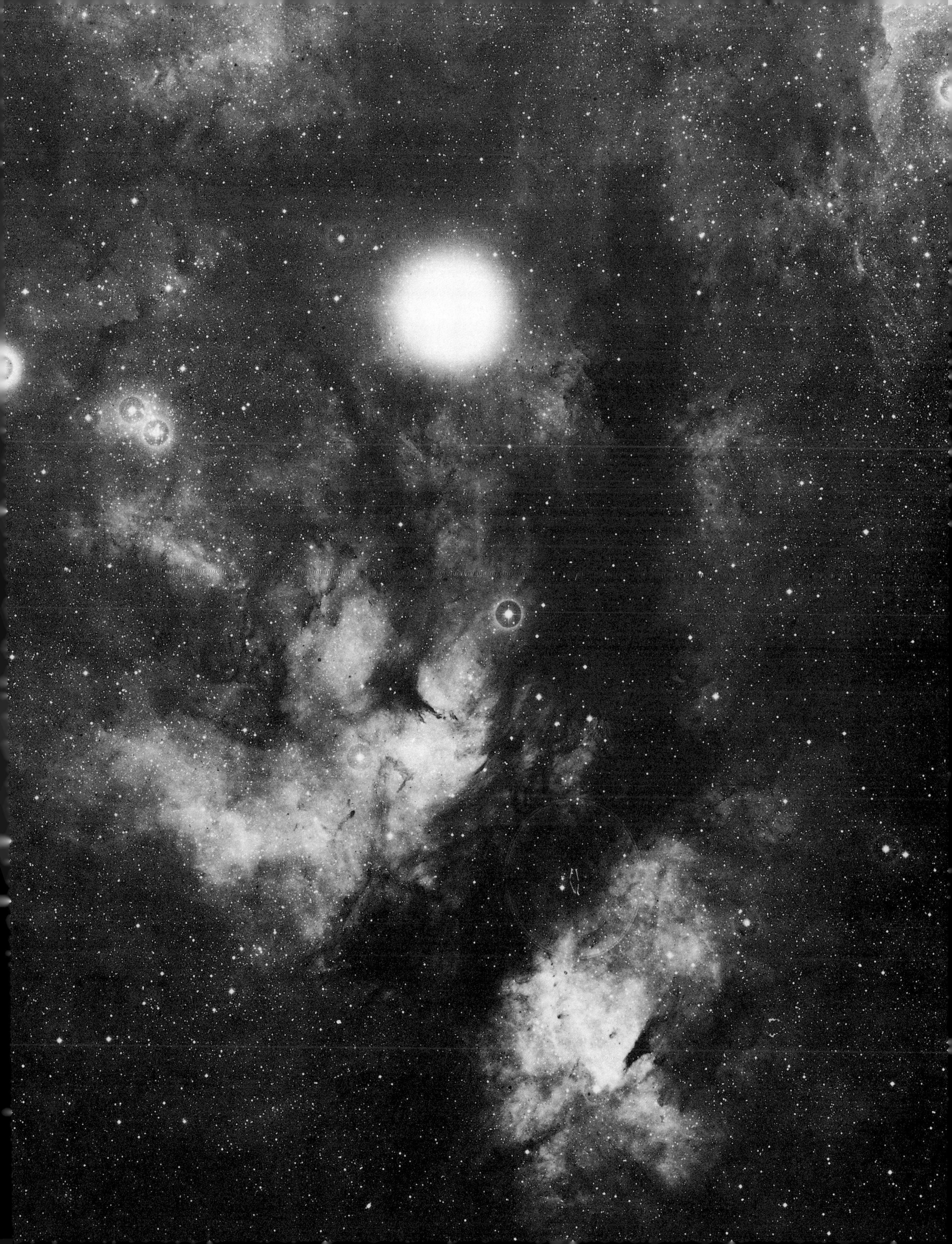

The Cygnus Loop

In Cygnus, there is a ring-shaped gaseous nebula with an angular diameter of 3°, which is impressive because of its regularity, its large size and its noticeably filamentary structure. The last feature clearly distinguishes it from diffuse gaseous nebulae which are generally chaotic in appearance. The object arose when the shock wave from an exploding star encountered the surrounding interstellar medium.

In the late evolutionary stages of massive stars an implosion caused by nuclear processes in the central core of the star may occur. During such an event, tremendous amounts of potential energy are transformed into kinetic energy and radiation. This causes the luminosity of such stars to rise to as much as 10^{10} times that of the Sun, and for most of the mass to be expelled into space. The star becomes a supernova. At least two principal types are recognized, governed by the nature of the object, and by the different physical processes that occur within them. On average, one supernova erupts in a galaxy every 30 to 50 years. In the area of the Milky Way that we can see, no more than five such events have been recorded in historical time. The number of known supernova remnants (about 150) is considerably higher than this figure. This is because the interaction of the intense flux of energy with the interstellar medium creates powerful radio and X-ray emission, and also because supernova remnants remain detectable for a very long time.

The Cygnus Loop is one of the best-known and most thoroughly investigated supernova remnants. Its visible filaments have been shown to have proper motions of about 0.03 seconds of arc per year across the sky, and spectroscopic examination reveals expansion speeds of slightly more than 100 km/s. This indicates a distance of 800 pc, a diameter of 40 pc and an age of over 50,000 years. In estimating these dimensions it had to be borne in mind that the gas immediately surrounding the star was first accelerated to velocities of the order of 10,000 km/s by the shock wave, but that then it underwent considerable braking caused by its continuous interaction with a resisting medium. These figures and the overall emission spectrum indicate that the Cygnus Loop is a relatively old supernova remnant.

The spherical shock wave expanding from the centre of the explosion produces a hot, compressed gas whose state can be described by a temperature of several million Kelvin.

When the compressed gas behind the shock-front cools, instabilities occur and the materials form filaments. This is connected with the fact that, in the denser gas, cooling processes are more effective and cause a rapid decrease in temperature. This, in itself gives rise to an even stronger increase in density and even faster cooling. Temperatures of several ten-thousand K are attained in the filaments and this is the reason for the optical emission from the gas. Measurements of the filaments extent in depth, and theoretical considerations indicate that the filaments are flat sheets of material instead of strings. The question of the long-term stability of such features remains unsolved, however.

The main part of the Cygnus Loop has a temperature that is far too high for its emission to be observed optically. Weak X-ray radiation between wavelengths of 1 nm and 6 nm was detected in 1970, opening up the way to make direct measurements. The radiation detector used then was carried by a rocket and, because of its low angular resolution, did not allow the object's shell structure to be detected. Since then, both high spatial resolution and measurements of the X-ray spectrum have become possible, as well as the detection of emission lines from highly ionized elements, such as that of oxygen, for example. The Cygnus Loop has also been intensely investigated at radio wavelengths. The emission's steep decline towards higher frequencies indicated the non-thermal nature of the radiation. It is synchrotron radiation emitted by fast electrons spiralling along lines of magnetic force. From measurements of the polarization of the radio waves it has been found that the magnetic field lines are directed along the filaments.

There is a good correlation between the spatial distribution of the emission observed at all the various wavelengths. In order to understand the processes in detail, it is necessary to consider the events that occurred as well as the distribution of non-stellar material that originally surrounded the supernova. When these factors are taken into account, the observed distribution of emission can be largely explained theoretically.

Fig. 6.37. The Cygnus Loop.

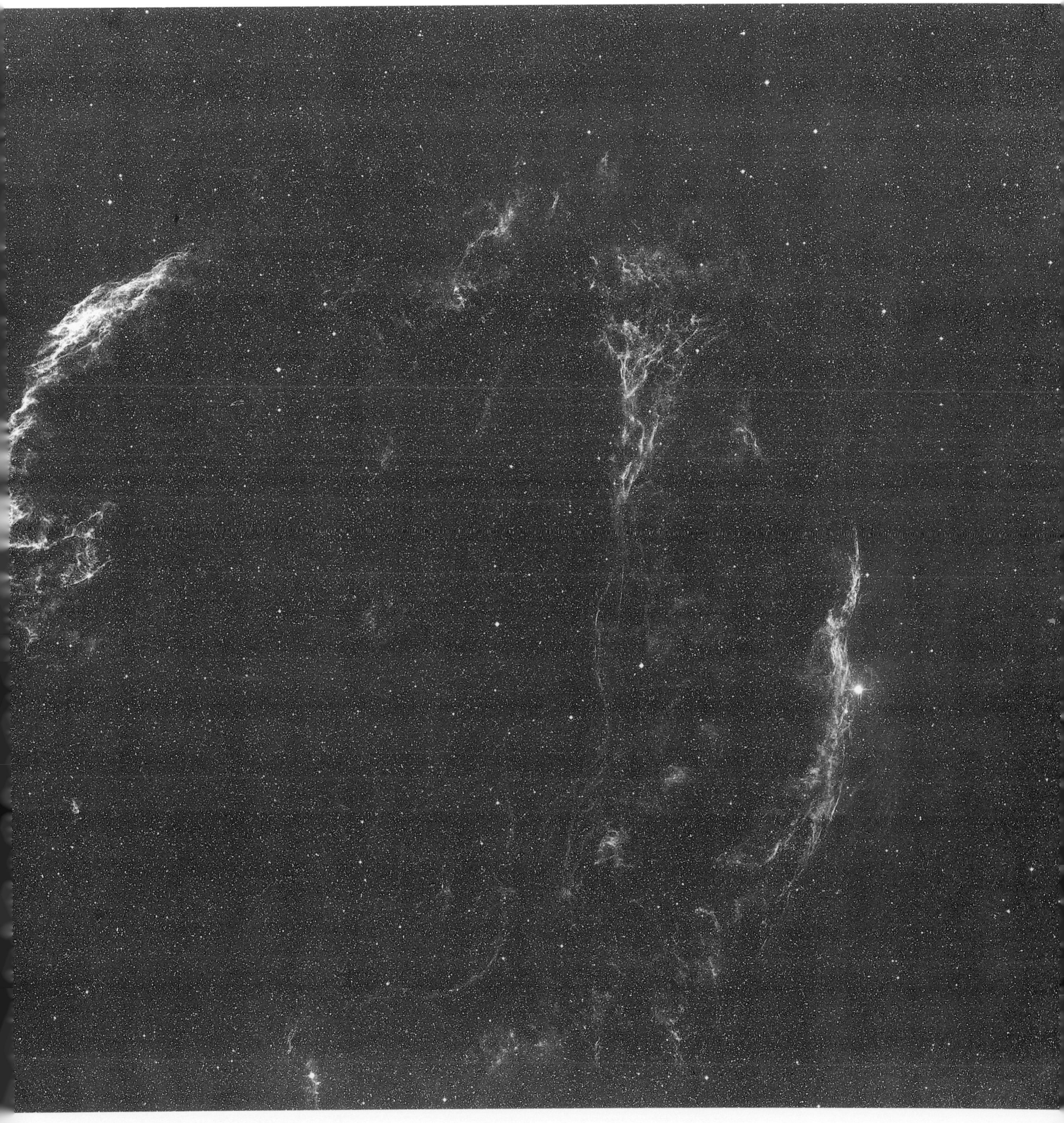

The Pelican Nebula

The western half of this picture is occupied by the Pelican Nebula in Cygnus. It is situated between the star α Cygni, not shown here, and the North America Nebula, which lies outside the eastern side of the picture. Because it is not very bright, the object is not included in the *New General Catalogue of Nebulae and Clusters of Stars* (NGC), which J. L. E. Dreyer published in 1888, and which was based on a catalogue by John Herschel. Its nomenclature comes from the two extensions of the NGC that were also published by Dreyer, in 1895 and 1908, and formed the *Index Catalogue* (IC). In this catalogue, two parts of the nebula are treated separately as IC 5067 and IC 5070, the latter being the one that forms the characteristic shape of a pelican.

Measurements at radio wavelengths show that the two parts, together with the North America Nebula and smaller areas of nebulosity form a complex of ionized gases – an extended H II region. The source of the ionization is presumably several stars with high surface temperatures, which, however, cannot be seen because of the amount of dust, except for one (HD 199 579). These interstellar dust clouds also partially absorb emission from the ionized gas at visible wavelengths and thus largely determine the structure of the overall nebular complex.

Areas of dense interstellar matter are generally also 'breeding grounds' for stars. The latter are formed as a result of gravitational instabilities, which can cause certain regions of suitably high densities and low temperatures to collapse. The potential energy released in this process causes a sharp rise in the temperature of the collapsing material, and at a certain stage, this, in turn, initiates fusion processes in the hydrogen gas: this productive source of nuclear energy turns the object into a true star. The specific evolutionary path is determined, both initially and later, by the mass of the star that is formed.

Various types of stars with every possible luminosity and mass have been recognized as being extremely young from specific peculiarities in their characteristics.

Very frequently, emission lines appear in the spectra, there is an excess of radiation in the infrared, and there are irregular changes in brightness. These are all signs of the existence of an unstable atmosphere that is the last phase in the process of star formation. We also know that interaction with the surroundings does not only involve accretion of material but also strong mass-loss from the star. Typical of these objects are the T-Tauri stars, named after the prototype star. Their masses are predominantly one solar mass or slightly less. T-Tauri stars occur in particularly large numbers in the region of the Pelican Nebula. They are a clear sign that stellar formation processes have occurred quite recently in this region. The oldest T-Tauri stars were formed there about a million years ago. In the course of the next 20 million years they will evolve into the stable state – known as the main sequence phase – of a normal star like the Sun.

Groups of young stars are described as associations, based on a concept suggested by the Soviet astrophysicist, Viktor Ambartsumian in 1946. When there is a concentration of T-Tauri stars, it is described as a T-Association. The Schmidt camera, with its large field of view is particularly suited to finding such T-Associations. Photographic plates obtained by such a telescope generally cover the entire area of the bright nebulosity and its surroundings. By comparing plates taken on different dates, changes in brightness – in this case of the T-Tauri stars – can be detected. Such observational contributions form an important basis for understanding where and how stellar-formation processes occur.

Fig. 6.38. The Pelican Nebula in Cygnus.

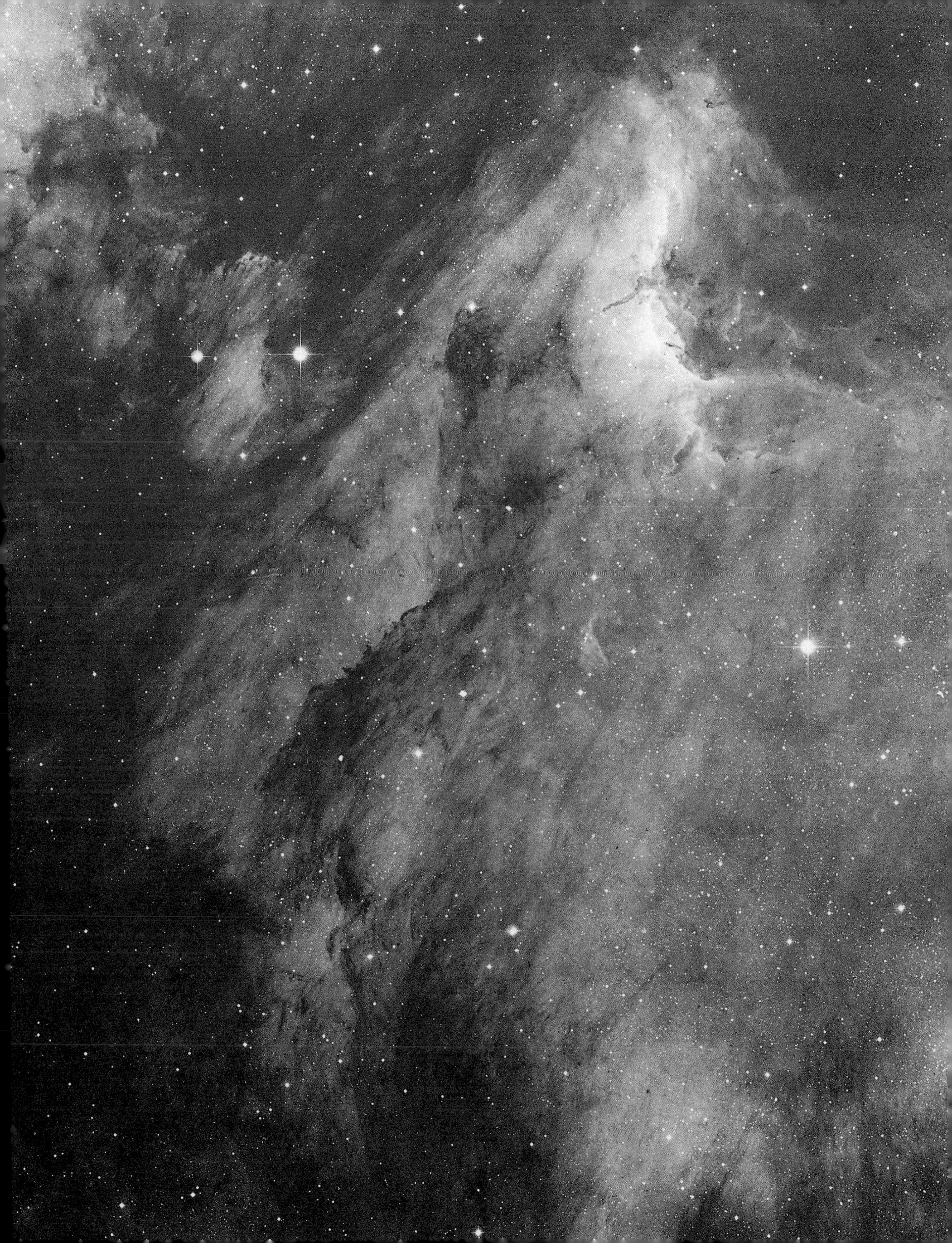

The North America Nebula: an original exposure taken by Bernhard Schmidt

This picture of the North America Nebula and its surroundings has been made from an original plate taken by Bernhard Schmidt at the Hamburg Observatory. The guide star used to control the guiding of the telescope during the two-hour exposure was ζ Cygni. The exposure was taken in September 1932, i.e. about two years after this ingenious optician had designed and built the first example of the special aplanatic system that was later named after him. The field covered was circular because during the exposure the plate was pressed onto a circular ring so that the emulsion would assume the same shape as the spherical focal surface.

This picture of this interesting area of the sky demonstrates convincingly the exceptional optical quality of the system developed by Bernhard Schmidt. An exceptionally large field of view of 7.5° angular diameter is combined with the highest image quality. Right out to the very edge of the plate, the images of stars do not show any aberrations, even when examined with a magnifying glass. The focal ratio of the optics in the 'original Schmidt' is also remarkable. With a corrector plate 36 cm in diameter and a focal length of 62.5 cm the ratio is *f*/1.75. This high light-grasp power is especially beneficial when photographing extended objects.

It is interesting to compare this exposure with the next, which was taken with the Schmidt telescope at the Karl-Schwarzschild Observatory in Tautenburg, the world's largest of this type. The free aperture is 134 cm, and the focal ratio is *f*/3.

Because of its large light-collecting surface the Tautenburg instrument is far superior to the original telescope in showing faint stars. In photographing extended objects, however, the focal length also plays a part of its effect on the plate scale. The smaller telescope is more effective because of its faster focal ratio.

If, despite this, the North America Nebula looks clearer on the exposure taken with the large Schmidt telescope, this can be attributed to the special printing process. This example shows that it is possible to extract more of the information content from a photographic plate by special treatment after the actual exposure. Better results can be obtained, as here, by a sophisticated darkroom technology and also by scanning the plate photoelectrically and applying digital image-processing techniques. With the Tautenburg plate, weak surface brightnesses were strongly enhanced by contrast control in the printing process. Details therefore become visible that were only latent on the untreated plate. The faint Pelican Nebula only appears clearly alongside the much brighter North America Nebula after this correction has been applied. Even around the star, ζ Cygni, near the 'West Coast' of the North America Nebula, thin bright veils can be recognized in places, whereas in Schmidt's untreated original, despite the light-grasp power of his telescope, a stellar void caused by interstellar extinction is particularly noticeable.

Another historic exposure of the North America Nebula was taken in 1972 by the American Apollo-16 Astronauts, using a Schmidt camera on the Moon. One of the scientific results returned by the astronauts were photographs of the Earth and other celestial objects in the ultraviolet. They took advantage of the Moon's lack of atmosphere, and obtained exposures in the spectral region below 160 nm, which is inaccessible from Earth. The aperture and focal length of this small Schmidt system were both 8 cm. It was used as a hand-held camera without any special mounting and could obtain either direct photographs or (with the use of an additional grating) spectra of objects. An interesting technical solution was employed so that the optical system served to image the ultraviolet radiation and, simultaneously, as an electronic image-intensifier.

Fig. 6.39. The North America Nebula in Cygnus. An original photograph by Bernhard Schmidt.

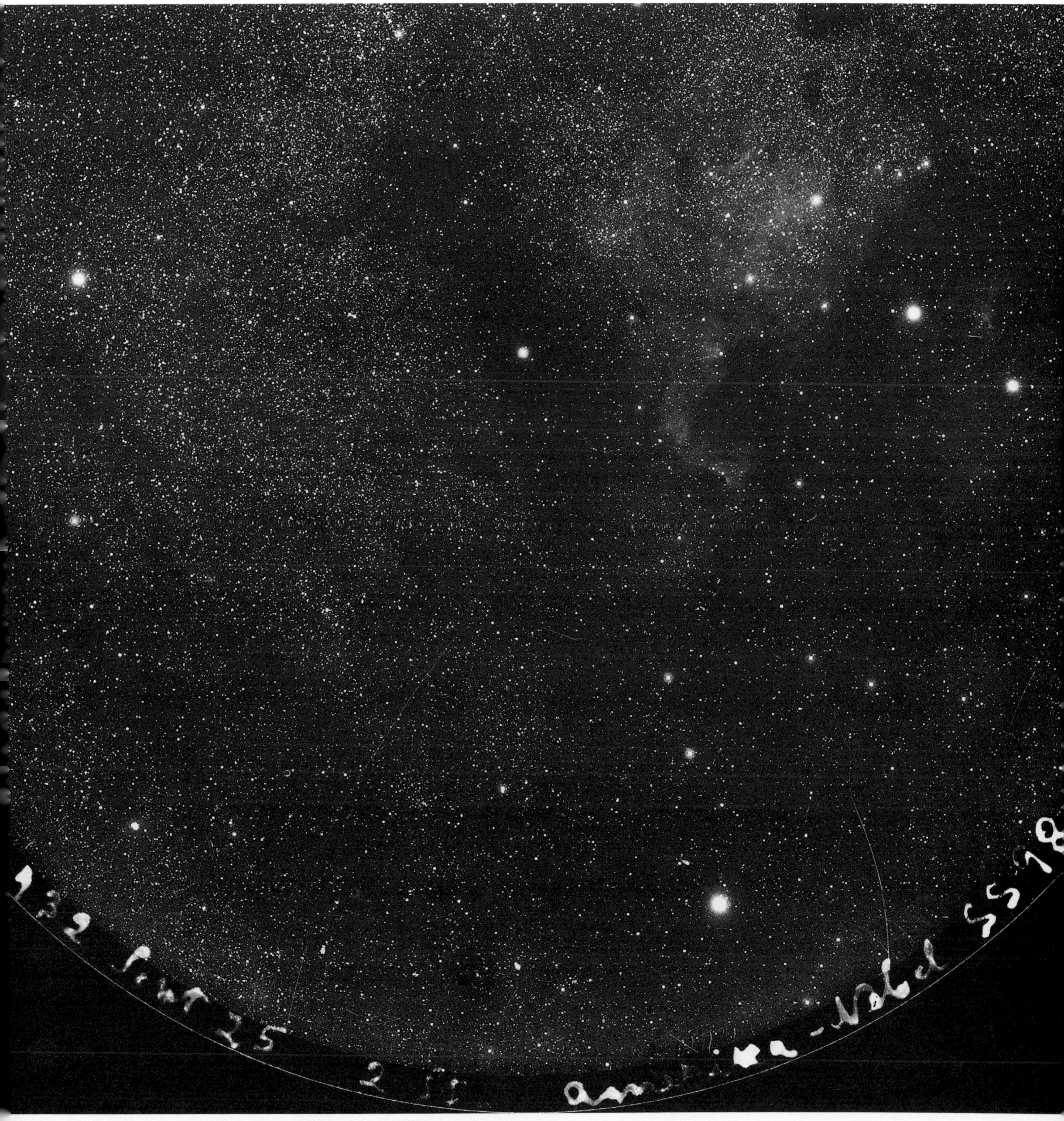

The North America Nebula (NGC 7000)

The North America Nebula lies some 3° east of α Cygni, the main star in Cygnus. The line of sight towards this area of the sky runs along the local spiral arm. This explains the unusually large number of stars and the frequent occurrence of various types of interstellar matter. The outline of the bright nebula on photographic plates has a great similarity with the North American continent, including the Florida peninsula and the narrow strip of land in Central America. Max Wolf, a pioneer of astrophotography, therefore suggested the name 'America Nebula' shortly after the turn of the century. Under especially favourable observing conditions the object can be seen with the naked eye, albeit as a very faint, luminous object that is less than 1° across. From his visual observations, John Herschel described the area as '... an immense nebulosity; too ill defined to fix the limits'. West of the North America Nebula there lies another bright nebula with a conspicuous shape: the Pelican Nebula (IC 5067 and IC 5070).

The shape of the North America Nebula stands out particularly clearly in the spectral lines of hydrogen. This shows that it is a gaseous nebula, and that the conspicuous shape is formed by absorbing clouds of dust. The latter are noticeable because of their lesser density of stars, as seen on the 'West Coast' or in the 'Gulf of Mexico'. The source of energy for the gaseous emission is the star HD 199 579, which can be found to the top right of the picture, vertically above the 'Florida Peninsula'. Its spectral energy distribution corresponds to spectral class 06 and it lies at a distance of 1200 pc. This value is a close approximation to the distance of the nebula. Other stars that might excite emission could be hidden behind dense local masses of dust. Evidence suggesting this has been gained by investigations in the near infrared region, where interstellar dust has little effect and long-wave radiation from stars can penetrate the dust clouds.

Analysis of all the observational data gives a linear diameter for the entire complex consisting of the North America Nebula and the Pelican Nebula of approximately 50 pc, and a total mass of about 20,000 solar masses, which corresponds to an average gas density of about ten protons per cubic centimetre. The complex is therefore an extended and evolved H II region.

The entire area is included, with the designation W80, as a radio source with a continuous spectrum in a catalogue compiled by the Dutch astronomer, G. Westerhout, in 1958 after one of the first comprehensive radio surveys of the sky. The continuous spectrum at radio wavelengths is caused by energy transitions of free electrons during encounters with protons, which are known as free–free transitions.

As mentioned on p. 140, the Apollo-16 astronauts obtained early ultraviolet exposures of the North America Nebula from the Moon in 1972. These plates indicated that the nebula is surprisingly bright at short wavelengths, but that it no longer bears any similarity to the North American continent. Indeed, the areas immediately surrounding several stars, including HD 199 579, emerge much more clearly.

Since then, it has been shown that this ultraviolet radiation is predominantly continuous emission, and it is interpreted as radiation from hot stars that is being very efficiently scattered by interstellar dust particles. The individual maxima in the intensity distribution of the ultraviolet radiation, on the other hand, lead to the conclusion that there are several energy sources that are responsible for ionization and emission in the large complex consisting of the North America and the Pelican Nebulae.

Fig. 6.40. The North America Nebula in Cygnus.

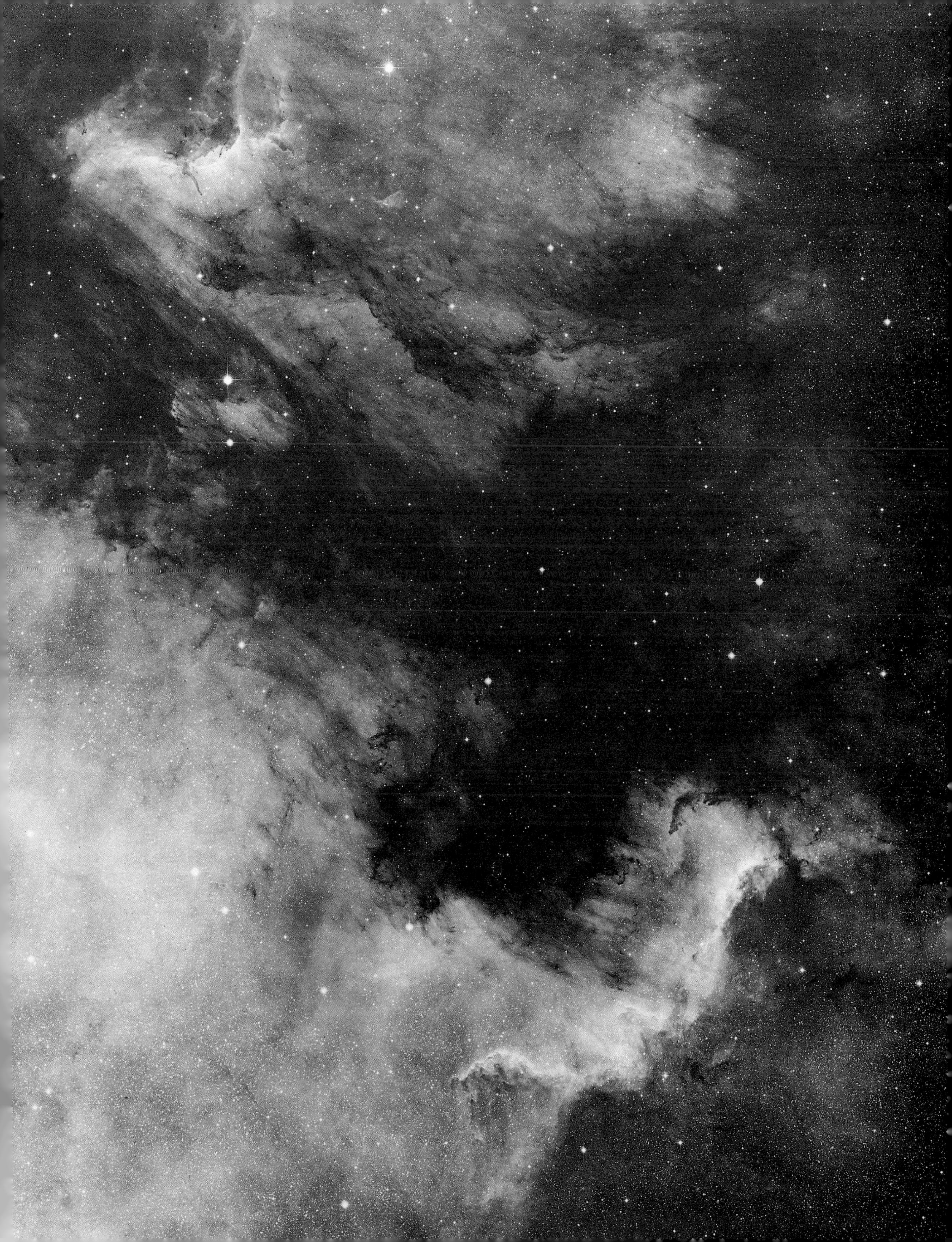

The Cocoon Nebula

The name of this small, bright nebula is derived from its overall appearance. A central star appears to be cocooned by the roughly circular shell of light, within which dark filaments can be seen. The angular diameter of the nebula measures only 12 minutes of arc, and it lies at the eastern end of a narrow dark cloud that has a length of nearly 2°. This is revealed by the number of stars that can be seen per unit area of the sky, which is much less than in surrounding areas. This is caused by interstellar dust which blocks the light from the distant stars that lie behind it. We can find this area in the region of the Milky Way that contains Cygnus, a region which is generally rich in various forms of interstellar matter.

In fact, there is not just a single star in the Cocoon Nebula: nearly all the stars situated within the area of the nebula are related to it. They are all members of the cluster IC 5146. By comparison, background stars are virtually invisible, because their light is absorbed by the dark cloud. Photometric investigations of the stellar radiation at different wavelengths show that the concentration of dust becomes strongest towards the centre, causing the central star to have an extinction of several magnitudes. IC 5146 lies at a distance of 1000 pc from us and is one of the very young galactic clusters, where the fainter members are still contracting on their way to the stable main-sequence phase. The object is an excellent example of the association of very young stars with gas and dust, which are the remnants of material left behind after the process of star formation.

Spectroscopic investigations show that the Cocoon Nebula is a bright gaseous nebula. Its spectrum contains strong emission lines, such as the Hα line of hydrogen at a wavelength of 656 nm. This indicates that quantum processes in ionized gases are the cause of the radiation: energy transitions in the electron shells of atoms of gas and ions after recombination, or after excitation by radiation or collisions, are responsible for the line emission. In this case, however, the line spectrum is superimposed on a stronger continuum, which can be attributed to scattering of starlight by interstellar particles occurring within the gas.

The obviously spherical shape of the Cocoon Nebula means that it is a typical example of what is known as a Strömgren sphere of ionized hydrogen gas. In 1939, Bengt Strömgren theoretically investigated the case where a star with a high surface temperature, embedded in interstellar matter, ionizes the surrounding gas with its high-energy radiation. Hydrogen should be particularly prominent because, as in the overall cosmic abundance of chemical elements, it also predominates in interstellar gas. Ultraviolet quanta with an energy of at least 13.6 eV are required to separate electrons from the hydrogen nuclei, the protons. Radiation quanta from the star encounter the surrounding gas and, provided that they are sufficiently energetic, cause ionization to occur. The spatial extent of the volume within which practically all the hydrogen is ionized, depends on the number of quanta that leave the surface of the star per unit time, on the density of the particles in the gas and on extinction caused by dust. In an ideal case, the spherical volume of a Strömgren sphere of ionized hydrogen is separated from the neutral gas by a boundary layer that is relatively thin in comparison with the overall diameter.

Only the brightest star in the cluster IC 5146 can be the source of excitation. It is of spectral class B1V. This means that it has a surface temperature of 22,000 K, and accordingly emits about 3×10^{45} ultraviolet quanta per second. In comparison, all the other members of the cluster make an insignificant contribution because of their lower luminosities.

Because of the quantum processes just mentioned and, in the Cocoon Nebula, the scattering of the starlight by particles of dust, the diffuse material becomes observable optically. In the ionized gas, however, the free electrons, which are moving in the field by the positive atomic nuclei, also emit continuum radiation at radio wavelengths. Measurement of this radiation and theoretical models can be used to estimate the mass of ionized gas. A value of 20 solar masses has been obtained in this way. Given the angular diameter and the distance, this corresponds to an average density in the Strömgren sphere of around 50 protons per cubic centimetre. By comparison, the mass of the surrounding neutral gas is about 30 times as great. This is also the result obtained from radio investigations. From the stellar population in the cluster, we can estimate that the total mass of stellar materials is of the same order as that of the neutral gas, i.e. between 500 and 1000 solar masses. The complex of gas and dust, in which the cluster IC 5146 was formed and a part of which is now ionized by the cluster, originally contained, therefore, about 2000 solar masses of material.

Fig. 6.41. The Cocoon Nebula in Cygnus.

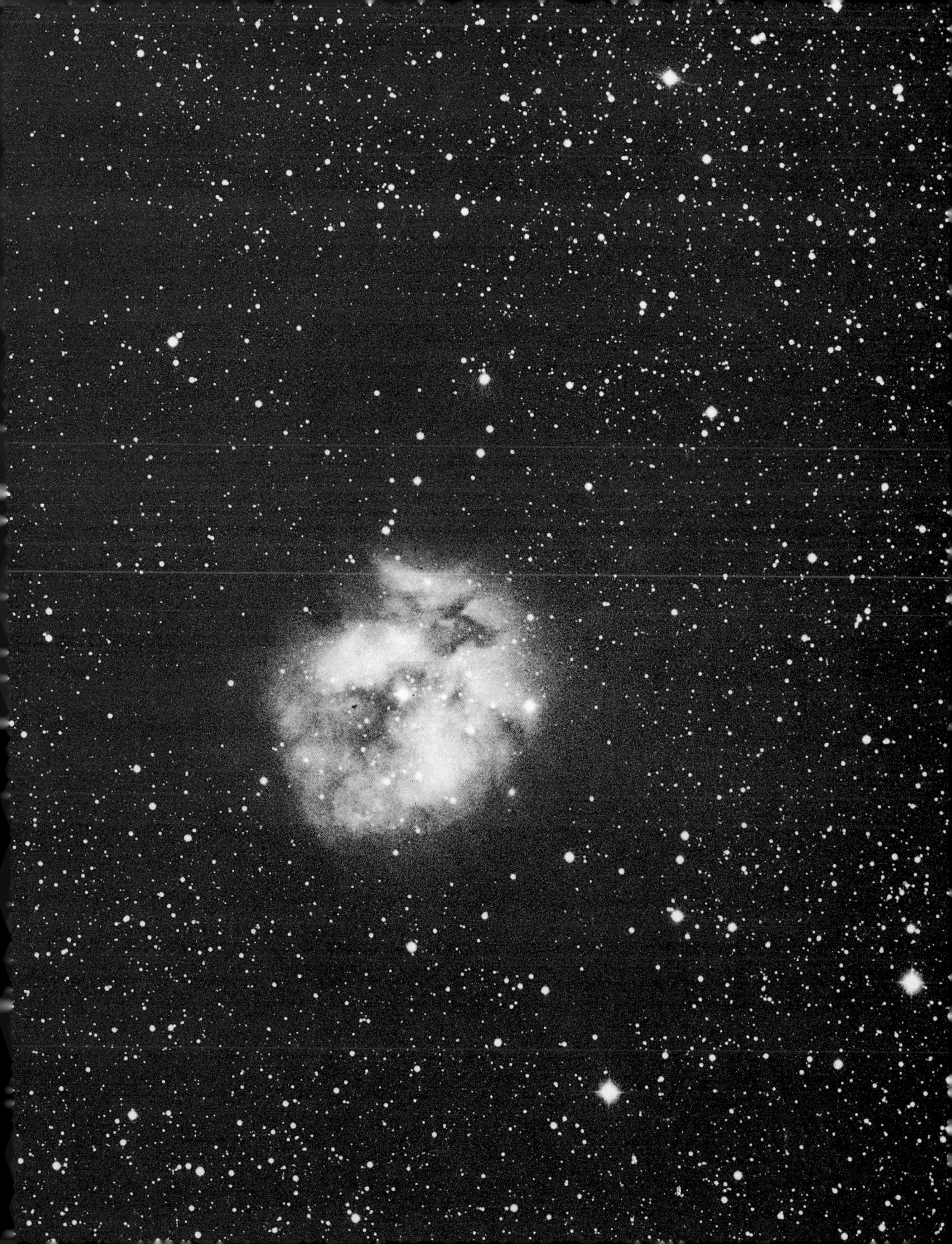

Star trails around the Celestial Pole

Even among newly built, large telescopes, mountings with the equatorial type of arrangement of the axes still tend to predominate. One of the two axes is directed towards the celestial pole, and it is known as the polar axis for this reason. Rotation of the telescope about this axis compensates for the change in position of stars over the course of a day. Such tracking – as it is called – is necessary for all astronomical observations in order to keep the telescope's field of view fixed relative to the sky during long observations or during the exposure of photographic plates on Schmidt equipment. This is achieved by a constant rotation of the polar axis. Steps have to be taken to ensure that the celestial pole and the instrumental pole always coincide. This is not only a question of convenience, but a basic prerequisite for the later interpretation of photographic plates taken by Schmidt telescopes. Constant monitoring with the guide telescope and appropriate minor corrections are required to ensure that the central star, the guide star, maintains exactly the same position on the plate and is correctly imaged. However, the field of view rotates about the centre if the positions of the two poles do not accurately coincide. This means that the stellar images will appear more deformed towards the outer edge, the longer the exposure lasts. Besides causing losses in limiting magnitude, this can lead to errors in subsequent photometric measurement of a plate. In aligning the polar axis with the celestial pole, either when first commissioning a new instrument, or when carrying out occasional checks, the deviation between the two directions has to be determined quantitatively by visual or photographic procedures. Appropriate mechanical adjustment of the polar axis can then be carried out accordingly with the adjusting screws. There will, however, be flexure of various parts of the telescope when pointing at different parts of the sky, because of the various directions in which the force of gravity is acting. This may mean that the position of the instrumental pole is dependent on the position of the telescope. The adjustment just carried out is therefore perfect for only one area of the sky and with long-exposure photographs a shift of the two poles relative to one another may be expected.

The photograph shown here was taken by the Schmidt telescope at the Karl-Schwarzschild Observatory for test purposes. It was taken to simulate an extremely long exposure and to serve to determine the shift between the true and the instrumental axes of rotation. The telescope was directed towards the celestial pole and the instrument was rotated quickly while the shutter was open. Each star left behind a closed track on the emulsion of an appropriate brightness. The most prominent trail is that of Polaris, which lies 45 minutes of arc from the true celestial pole. In order to obtain positional information, the rotation was interrupted every 15°, which produced bright markers on the artificial trail of each star. During the rotation of the telescope, the stars describe portions of circular arcs around the instantaneous axis of rotation. If this shifts because of flexure in the telescope, then this will also lead to a slight displacement of the photographic plate in relation to the sky. The star trails, which ideally would be circular, are deformed because portions of circular arcs of varying radii – i.e. at varying distances from the axis of rotation – are combined on the plate. This can be seen on the photograph shown here. Significant deviations from the circular shape can be identified, particularly with the faint objects near the pole, but also on the neighbouring star trails further towards the outside. This leads to the conclusion that, even with this telescope, there is a shift in the direction of the pole during rotation around its polar axis.

If the trails of different stars are analysed mathematically using the hour markers, then the behaviour of the axis of rotation at different orientations of the telescope can be determined graphically or arithmetically. For the telescope at the Karl Schwarzschild Observatory, this shows that the maximum flexure between opposite positions is two minutes of arc. Compared with other telescopes of similar design, this may be regarded as very satisfactory. In practice, the amount of shift does not influence measurements, and speaks well for the telescope itself.

Fig. 6.42. Star trails on a plate of the North Celestial Pole.

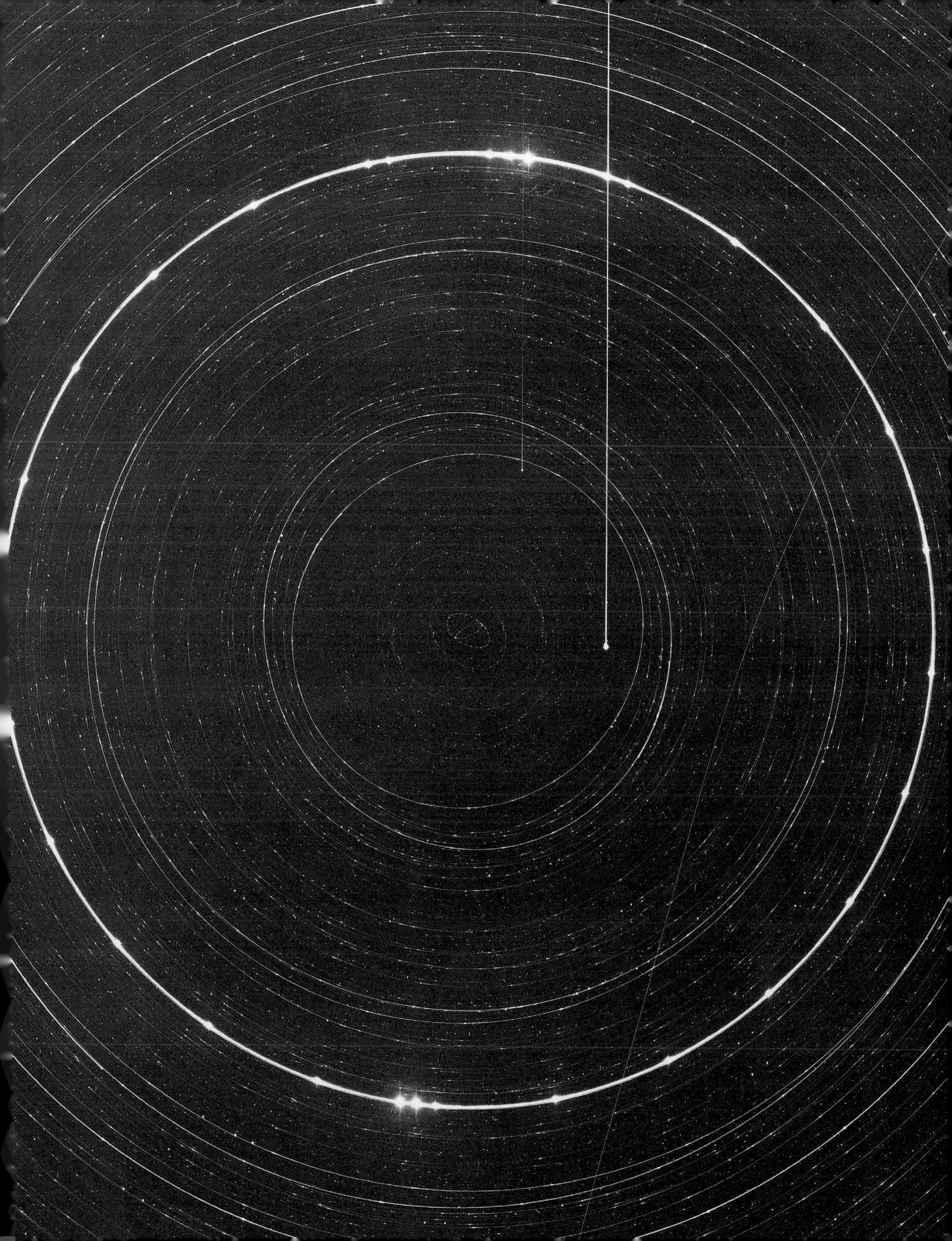

Comet Bennett

Comet Bennett was one of the brightest comets visible with the naked eye in the last few decades. Comets pose many unanswered questions for science, and they are also of particular interest to everyone, because of their often striking appearance. As a result, Comet Bennett was one of the most thoroughly investigated celestial objects.

This comet was discovered by a very active amateur astronomer, John C. Bennett on the 28 December 1969, as an object of magnitude 8.5. Bennett himself wrote that he achieved this success after 153 hours of systematic searching for comets in 1969 alone, and 333 hours in total. The comet he discovered is one of the most interesting. After only four days of observation, an orbit was calculated for Comet 1969i. This classification means that Bennett's Comet was the ninth comet discovered in 1969. The orbit's inclination of 90.3° meant that it was almost at right angles to the ecliptic and that the perihelion passage should occur on 20 March 1970 at a small perihelion distance of only 0.54 AU. Because of this close proximity to the Sun, the comet's apparent magnitude was expected to be very high.

In fact, Comet Bennett reached magnitude 0 between 20 and 25 March 1970, and the size of the tail visible to the naked eye amounted to 20°. It is therefore not surprising that Comet Bennett has often been compared to the historically significant apparitions of comets in 1858 (Comet Donati) and 1910 (Comet Halley).

The few photographs of the comet that can be shown here give only a slight indication of the variety in its appearance. Figures 6.43 and 6.44 were taken at an interval of a few days, and reveal significant differences in the structure of the tail. Figures 6.45 and 6.46 were taken, one immediately after the other, but in different spectral regions – one in the ultraviolet and the other in the red. Hundreds of photographs would have to be taken to show all the structures and changes in the tail of the comet.

Like many other comets, Comet Bennett was observed to have two tails. One consists of gas and; in accordance with the classification proposed by the Russian astronomer, T. Bredikhin, is known as a Type I tail. It is narrow and straight in shape. A type II tail consists of dust, and in the case of Comet 1969i it was bright, wide, and slightly curved. The Type I tail showed many rapid changes in structure, whereas the Type II tail was much more constant in its appearance.

Comets form a relatively small group of celestial objects. By 1950, about 800 had been discovered, and the number has still not passed 1000. Astronomy has only been concerned with comets for about 400 years. Even in the sixteenth century, it was thought that comets were emanations from the Earth, and that the rising gases caught fire in the Earth's atmosphere. From observations of a comet in the year 1577, Tycho Brahe was able to prove that its distance was greater than that of the Moon and that comets were therefore astronomical objects. Orbital calculations by Edmond Halley, involving the comet that was later named after him, then showed that there are periodic comets, i.e. that there are comets that follow elliptical orbits around the Sun, like the planets. 'Why are we so surprised that a rare apparition in the heavens does not appear to be governed by any laws, and its beginning and end remain unknown to us, when it only returns after great intervals of time? The moment will come when the course of time and centuries of research will bring to light things that are still hidden. One human lifetime is not enough for the investigation of such immense natural phenomena. At some time, the man will be born who will be able to show the position of cometary orbits, the reason their motion differs from that of the other heavenly bodies, their sizes, and their nature.'

Nearly 2000 years have passed since Seneca's far-sighted words. Has this been enough time to bring to light everything that was hidden? How do things stand, for example with cometary orbits, which Seneca also mentioned? From observations of about 600 comets, it can be seen that 40% of them, just like the planets, describe elliptical orbits around the Sun. With the planets these ellipses are very nearly circles – i.e. the eccentricity, the deviation from a circle – is very low. The elliptical orbits of the comets are very eccentric, they deviate greatly from a circle, and the value of the eccentricity is close to 1. A curve with an eccentricity of 1 is a parabola. This, in contrast to a circle or an ellipse, is not closed, nor is a hyperbola, whose eccentricity is greater than one. Parabolic orbits have been observed in about 50% of comets, and hyperbolic orbits in about 10%.

Objects that really do move in a parabolic or hyperbolic orbit come from space, swing round the Sun and then disappear back into space. Only those bodies with elliptical orbits are really bound to the Sun. International Astronomical Union Circular No. 2219 for 27 February 1970 gives the eccentricity of Comet Bennett as 0.996,000, and remarks that the deviation from the parabolic orbit is significant. Comet Bennett is therefore one of those objects with elliptical orbits around the Sun.

As they travel along their orbits, comets are very likely to be influenced by the gravitationnal fields of the large planets. The slightest perturbations can alter the course of their orbits. The hyperbolic orbits of 21 comets were examined, and it was found that 20 of them had been created by perturbations from originally elliptical orbits.

Fig. 6.43. Bennett's Comet taken in the ultraviolet.

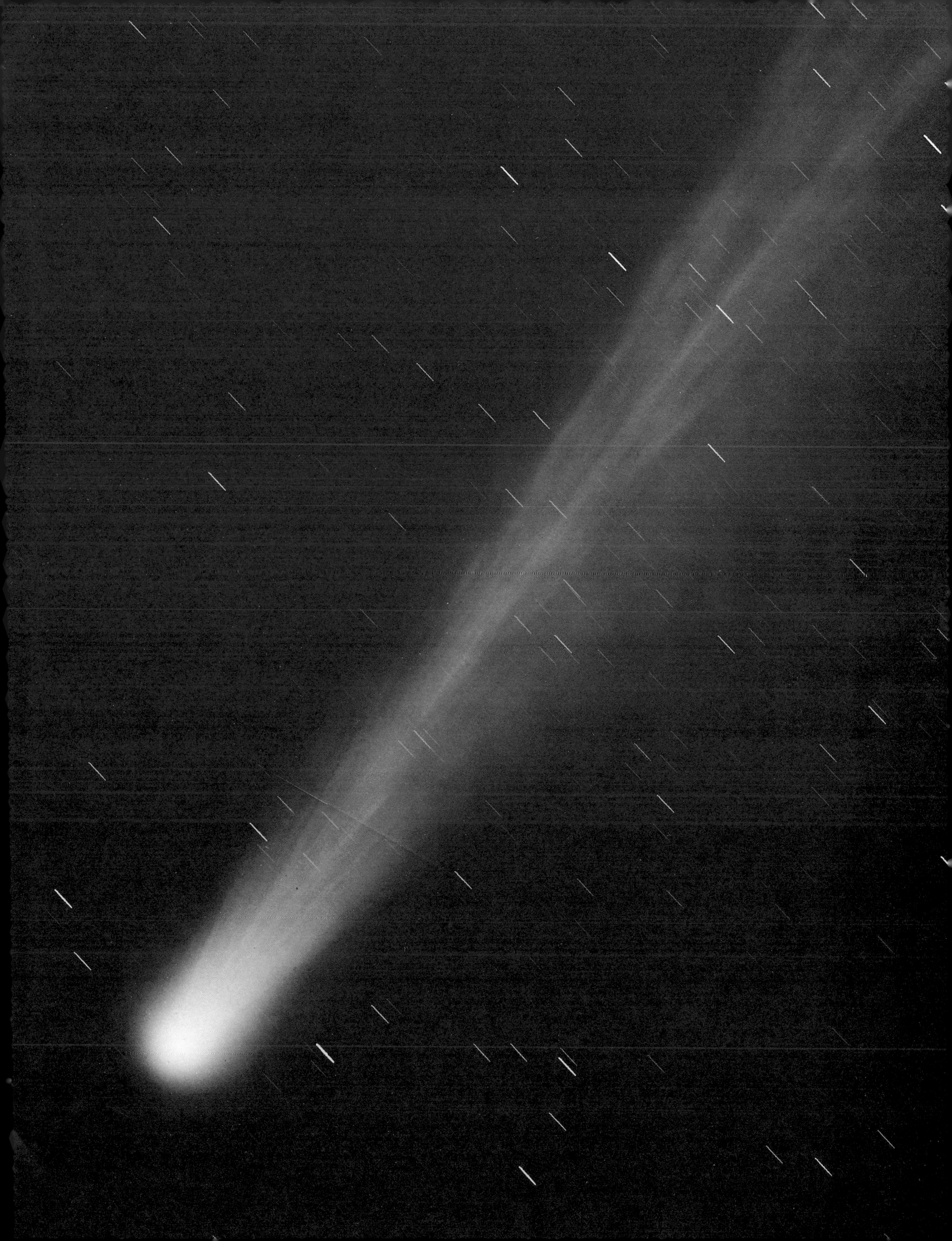

We can therefore assume that the majority of comets follow long elliptical orbits and are thus members of our Solar System. This example, however, also indicates that even in the field of cometary orbits there are still a lot of questions that need to be answered.

The motion of the comets cannot be examined independently of their appearance, size and nature. As planets can cause significant perturbations in the orbits of comets without being affected themselves, we can conclude that the mass of a comet is far smaller than that of a planet.

When looking at the pictures of Comet Bennett taken at short wavelengths, we can immediately see that the comet consists of a head and a tail. The tail only appears, however, when the comet comes within 2 AU of the Sun. Otherwise all that can be seen is the diffuse head of the comet. This also increases in size the nearer the comet approaches to the Sun. As the tail always points away from the Sun, it is assumed that the Sun causes the formation of the tail.

How then can these observations and the nature of comets be explained? It is generally accepted nowadays that, at a great distance from the Sun, the actual body of a comet is a solid nucleus, consisting of small dust particles embedded in frozen gas. Until recently it had been assumed that the diameter of a comet's nucleus amounted to between 5 and 10 km. These sizes were calculated from the following considerations. Cometary nuclei are visible at a great distance from the Sun, because they reflect light from the Sun. The intensity of the reflected light may be measured, and if we make certain assumptions about the reflectivity we can calculate the size of the reflecting surface. On the assumption that cometary nuclei reflected about 25% of the incident radiation, their diameters were thought to be between 5 and 10 km.

The diameter of the nucleus of the famous Comet Halley has always given about 6 km as derived by this method. Direct observations and measurements made by ESA's Giotto spacecraft and the Soviet VEGA probes show that the nucleus of Comet Halley is similar to the frustrum of a cone with a large diameter of 8 km, a small diameter of 5 km, and a height of 16 km. A great surprise was the fact that the nucleus of Comet Halley does not reflect 25% of the incident light but only 4%. If the 'mirror' reflects less, the reflecting surface has to be larger in order to explain the observed brightness. The false assumption about the reflectivity, meant that Comet Halley was given too small a diameter.

If the results gained for Comet Halley also hold for Comet Bennett, its nucleus would be about 20 km across, but would still not be measurable from the Earth. At a distance of 2 AU 20 km subtends an angle of only 0.014 seconds of arc. Even at 0.5 AU, the minimum distance reached by Comet Bennett, this would amount to only 0.055 seconds of arc. These angles are below the resolving power of any optical telescope.

When a cometary nucleus – which consists of frozen gases enclosing small solid particles – approaches the Sun on its elliptical orbit, it enters regions of increasingly stronger radiation. This leads to evaporation of the frozen gases. Because the mass of a comet's nucleus is small, its gravity is also low, and the evaporated gases can form an extended gaseous atmosphere, the coma, around the nucleus of the comet. Dependent on the amount of gas a coma may attain a diameter of between 10,000 and 100,000 km.

Spectroscopic examinations of the coma show both a continuous Fraunhofer spectrum, and emission lines and bands of various atoms and molecules. The continuous spectrum indicates that dust particles in the coma scatter light from the Sun. The emission lines come from carbon molecules, CN, NH, CH and OH compounds, and iron, nickel and sodium atoms. The molecules and atoms are excited into fluorescence by solar, energetic short-wave radiation. The atoms and molecules absorb the Sun's radiation and then re-emit it as emission lines.

The density of the coma around a cometary nucleus is very low. Near the nucleus it amounts to approximately 10^{13} molecules per cubic centimetre, and at the outer edge only about 1000 molecules per cubic centimetre. Compared with this, a cubic centimetre of air at the surface of the Earth contains 3×10^{19} molecules. Even near the nucleus of the comet the density of the coma amounts to only 10^{-6} of the density of the Earth's atmosphere.

As mentioned earlier, a comet often has two tails. The first is directed almost radially away from the Sun, and the other curves slightly away from this direction. Spectroscopic examinations have shown that the radial, Type I tail, which may attain a length of 10^8 km, consists of gas. Emissions from CO and CO_2, and from HC, N and OH compounds, have been observed from the tail. In contrast to the coma, the molecules in the tail are ionized. The density of the molecules in the tail is even lower than in the coma, and amounts to about 100 molecules per cubic centimetre.

The slightly curved, Type II tail consists of dust-like material, and this can be recognized from its continuous spectrum. Dust tails are generally shorter than gas tails, are less luminous and show less structure.

The exceptionally low density of the tails is also clear in the photographs of Comet Bennett because numerous faint stars can still be seen through the tails.

If the force responsible for the tails comes from the Sun, then it must be more powerful than the Sun's gravitational attraction on the material in the tail. As early as 1943 Cuno Hoffmeister suggested that the direction of the Type I tail was determined by corpuscular radiation from the Sun. Ludwig Biermann then postulated a continuous solar wind, which expels the gas from the coma. Space probes later discovered the solar wind, which consists of electrons, protons and occasional helium nucleii, all of which are charged particles. The velocity of the solar wind at a distance of 1 AU from the Sun is between 400 and 500 km/s. Its average density is about 6–8

Fig. 6.44. Bennett's Comet taken in the ultraviolet.

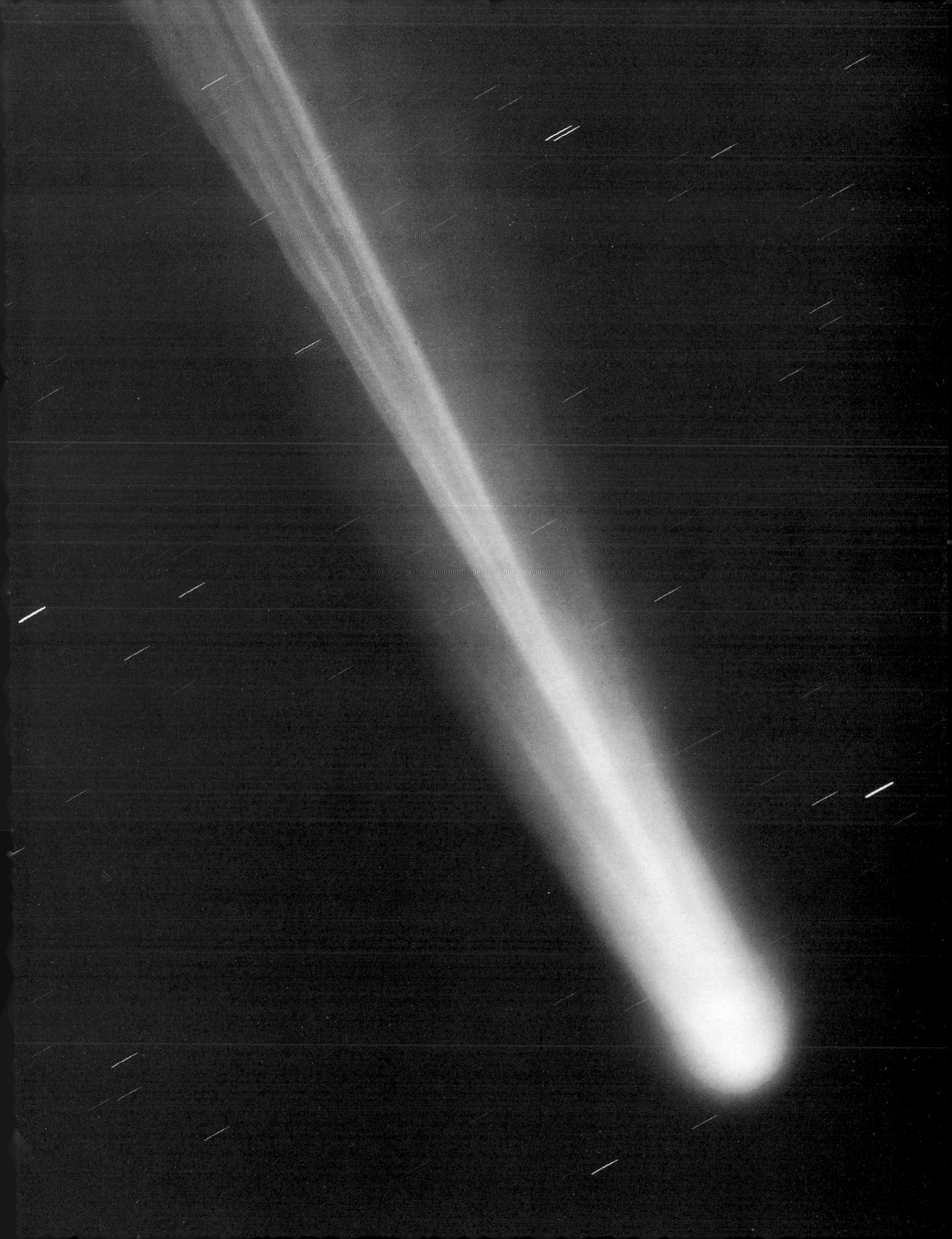

particles per cubic centimetre. There are, however, considerable changes in the strength of the solar wind, and the particles are often ejected from the Sun in the form of high-speed streams. This partly explains the frequently highly variable features found in the cometary tails, as can be seen in the pictures of Comet Bennett.

The solar wind does not have any effect on the dust particles. These are forced out of the coma by radiation pressure from the Sun. The effect of radiation pressure depends on the size of the particles, and this leads to a broadening of the dust tail. Because the dust particles leave the coma with a velocity considerably lower than that of the gaseous components, the direction of the dust tail is not only determined by the speed of their emission from the coma, but also by the orbital velocity of the comet, and this leads to the observed curvature.

As the comet recedes from the Sun, first the tail, and then the coma disappears. Each time a comet passes the Sun it loses part of its mass because of the formation of the tail. Estimates give a loss of the order of 10^{-3} of the total mass per orbit. This leads to the conclusion that the life of a comet is extremely limited. It is possible that we live just in that phase in the evolution of the Solar System when comets occur, but there is no evidence to support this rather improbable explanation. Therefore, there has to be some sort of reservoir from which the comets that we observe are produced. According to J. Oort, this reservoir appears to be a giant cloud, which surrounds the Sun at distances of up to 50,000 AU, and which contains about 10^{11} cometary nuclei. These nuclei are thought to be remnants of an early phase in the formation of the Solar System. As a result of mutual perturbations and the influence of neighbouring stars, individual cometary nuclei are constantly directed in towards the inner regions of the solar system and are thus trapped by the Sun. The figure of 10^{11} cometary nuclei might appear to be very high. When we consider, however, that the mass of one nucleus amounts to about 10^{17} or 10^{18} g, then the total mass of the nuclei amounts to only ten times that of the Earth.

Much light has been shed on the problems raised by comets since the time of Seneca, and an important part in this has also been played by Comet Bennett. There are, however, still many questions that remain to be answered.

Fig. 6.45/6.46. Bennett's Comet taken in the ultraviolet (right) and in the red (left).

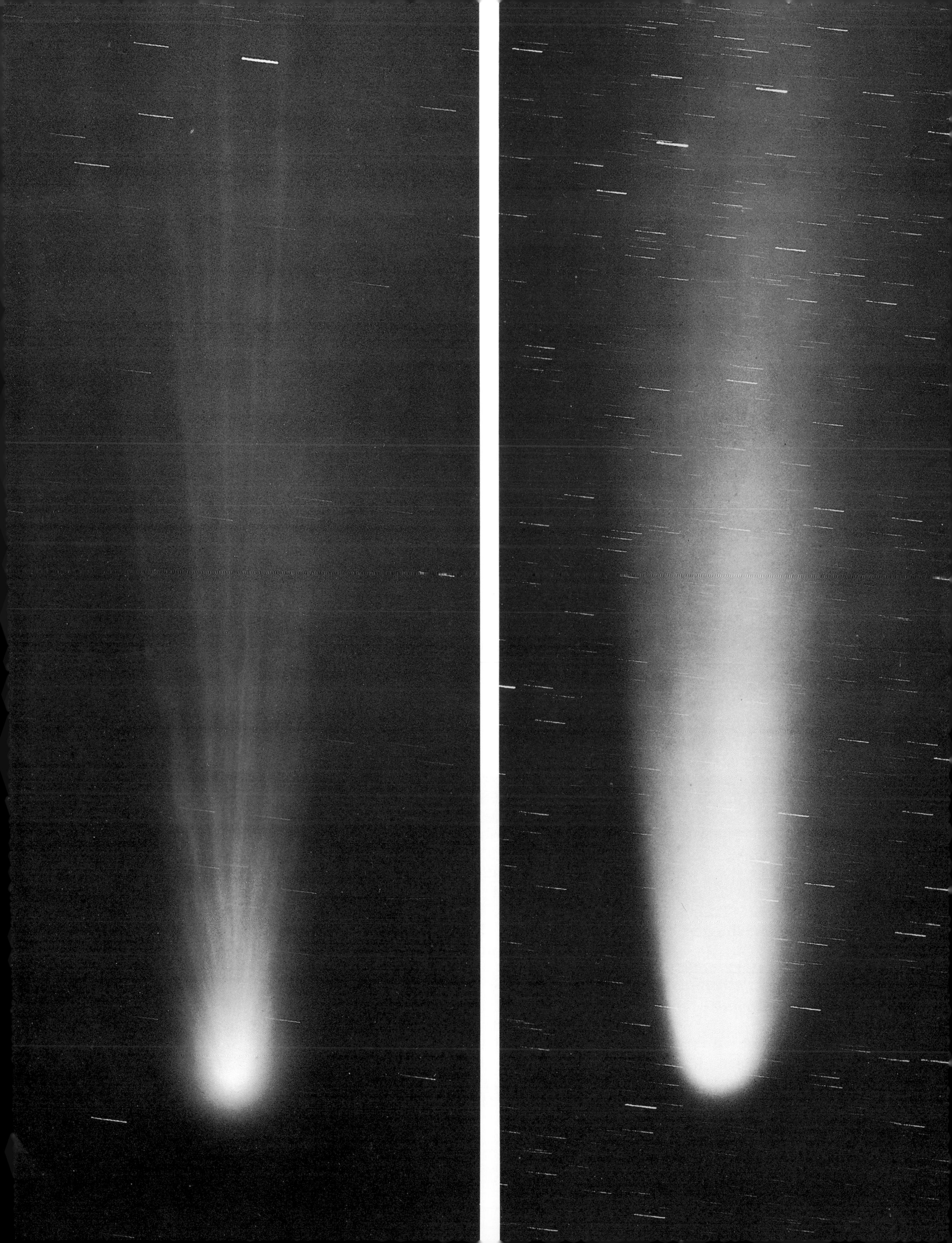

References

Ambronn, L. (1899). *Handbuch der Astronomischen Instrumentenkunde*, 2 Bde. Berlin.

Beatty, J. K. (1985). HST – Astronomy's Greatest Gambit. *Sky and Telescope*, **69**, 409.

Beck, R. & Wielebinski, R. (1981). Radio Waves from M31 and M33. *Sky and Telescope*, **61**, 495.

Bertola, F. (1983). Die große Magellansche Wolke. *Sterne und Weltr.*, **22**, 115.

Bertola, F. (1984). M82 – NGC 3034. *Sterne und Weltr.*, **23**, 568.

Bertola, F. (1984). NGC 5184/NGC 5195. *Sterne und Weltr.*, **23**, 622.

Biermann, P. (1985). Explosionsartige Sternentstehung in der irregulären Galaxie M82. *Sterne und Weltr.*, **24**, 202.

Block, D. L. (1986). How Big is M81? *Sky and Telescope*, **72**, 454.

Börner, G. (1984). Das Alter der Kugelsternhaufen. *Sterne und Weltr.*, **23**, 422.

Brauer, R. (1983). Die Überschalldüsen von Galaxien. *Sterne und Weltr.*, **22**, 173.

Brinkmann, W. (1985). Der Krebs-Nebel – ein Prüfstein für Supernovavorstellungen. *Sterne und Weltr.*, **24**, 76.

Capaccioli, M. (Ed.) (1984). *Proc. 78th IAU Coll. on Astronomy with Schmidt Telescopes.* Dordrecht.

Carruthers, G. R. & Page, Th. (1972). The UV Schmidt Camera Experiment Aboard Apollo 16. *Sky and Telescope*, **43**, 151.

Cederblad, S. (1946). Studies of Bright Diffuse Galactic Nebulae. *Medd. Lund Obs.*, Ser. II, No. 119.

Chincarini, G. & Rood, H. J. (1970). The Cosmic Tapestry. *Sky and Telescope*, **59**.

Curtis, H. D. (1918). The Planetary Nebulae. *Publ. Lick Obs.*, **13**, 57.

Dufour, R. J. & van der Bergh, S. (1978). Structure and Evolution of NGC 5128. *Sky and Telescope*, **56**, 389.

Elsässer, H. (1985). *Weltall im Wandel – Die neue Astronomie.* Stuttgart.

Felli, M., Churchwell, E. & Massi, M. (1984). A High Resolution Study of M17 at 1.3, 2, 6, and 21 cm. *Astron. Astrophys.*, **135**, 53.

Gieseking, F. (1985). Die Zentralsterne Planetarischer Nebel. *Sterne und Weltr.*, **24**, 577.

Goudis, C. (1982). *The Orion Complex – A Case Study of Interstellar Matter.* Dordrecht.

Hale, G. E. (1935). The Astrophysical Observatory of the California Institute of Technology. *Astrophys. Journ.*, **82**, 111.

Hall, D. N. B. (Ed.) (1982). Proc. of Special Session of IAU Comm. 44 on the Space Telescope Observatory. *Publ. Space Tel. Science Inst.*

Harrington, R. G. (1952). The 48-Inch Schmidt Type Telescope at Palomar Observatory. *Publ. Astron. Soc. Pacific*, **64**, 275.

Herbig, G. H. (1968). The Structure and Spectrum of R Monocerotis. *Astrophys. Journ.*, **152**, 439.

Hodapp, K.-W. (1982). Die Entstehung von Galaxienhaufen. *Sterne und Weltr.*, **21**, 285.

Hodges, P. C. (1953). Bernhard Schmidt and His Reflector Camera, in *Amateur Telescope Making*, Book III, ed. Ingalls, A. G., Scientific American, New York.

Hopp, U. (1986). Entdeckung von Delta-Cephei-Sternen in M101. *Sterne und Weltr.*, **25**, 571.

Kasten, V. (1982). Gibt es die Oortische Kometenwolke wirklich? *Sterne und Weltr.*, **21**, 196.

König, A. (1937). *Die Fernrohre und Entfernungsmesser.* 2 Aufl. Berlin.

Kopal, Z. (1956). Astronomy and Optics. In *Astronomical Optics and Related Subjects.* Amsterdam.

Lassell, W. (1865). Brief an den Herausgeber. *Astron. Nachr.*, **63**, 369.

Malin, D. & Murdin, P. (1986). *Colours of the Stars*, Cambridge.

Meisenheimer, K. (1982). Die Verteilung der galaktischen Kugelsternhaufen. *Sterne und Weltr.*, **21**, 349.

Meisenheimer, K. (1982). Das Rätsel der Jets. *Sterne und Weltr.*, **21**, 462.

Merkle, F. & Schneermann, M. (1986). Das ESO Very Large Telescope. *Sterne und Weltr.*, **25**, 460.

Mitton, S. (1977). *The Cambridge Encyclopaedia of Astronomy.* New York.

Mürsepp, P. W. & Weismann, U. K. (1984). *Bernhard Schmidt.* Nauka.

Oort, J. H. (1982). Die Entwicklung großräumiger Strukturen im Universum. *Sterne und Weltr.*, **21**, 456.

Rahe, J. (1977). Staubproduktion von Kometen. *Sterne und Weltr.*, **16**, 47.

Riekher, R. (1957). *Fernrohre und ihre Meister.* Berlin.

Robinson, L. (1982). An Eye for Tomorrow. *Sky and Telescope*, **63**, 128.

Robinson, L. (1986). Update – Telescopes of the Future. *Sky and Telescope*, **72**, 23.

Schaifers, K. & Traving, G. (1973). *Meyers Handbuch über das Weltall.* 5. Aufl. Mannheim.

Smith, D. H. (1985). Mysteries of Cosmic Jets. *Sky and Telescope*, **69**, 213.

Steshenko, N. V. (1979). On the Feasibility of the 25 Meter Optical Telescope. *Smithsonian Astrophys. Obs.* Special Report No. 385, p. 191.

Trefzger, Ch. (1978). Chemische Entwicklung von Populationen und Galaxien. *Sterne und Weltr.*, **17**, 14.

Tucker, W. (1985). The Space Telescope Science Institute. *Sky and Telescope*, **69**, 295.

Van den Bergh, S. & Dufour, R. J. (1978). Farbbeobachtungen von NGC 5128=Centaurus A. *Sterne und Weltr.*, **17**, 19.

Wätzig, A. & Zeiger, G. (1985). *Bernhard Schmidt – eine biographische Skizze.* Wissenschaftl. Arbeiten Ing-Hochschule Mittweida (Sonderheft).

Weigert, A. & Zimmermann, H. (1977). *Brockhaus ABC Astronomie.* 5 Aufl. Leipzig.

Weinberger, R. (1976). Die lokale Gruppe von Galaxien. *Sterne und Weltr.*, **15**, 388.

Wilson, R. N. (1982). Progress on the 3.5-m New Technology Telescope (NTT). *ESO Messenger* No. 29, p. 24.

Wilson, R. N. (1986). Active Control and Mirror Support for the VLT. *Proc. 2nd ESO Workshop.* München.

Wolff, S. C. (1985). A Selective History of Telescopes. *Mercury*, **14**, 139.

Appendix

A fast, coma-free reflecting system* [with Plates I and II]

A translation of the text of Bernhard Schmidt's only scientific paper on the Schmidt telescope design (pp. 62–3).

If the light-loss in a reflecting and a refracting system are compared, it is found that, for equal aperture ratios, the reflector shows a smaller light-loss than the refractor. A freshly silvered mirror reflects at least 90% of the incident light, whereas doublet or triplet systems transmit at the most 80% or 70%, respectively. The situation becomes even less favourable with the larger lenses because of the increasing absorption of short-wave light.

In large telescopes, therefore, a parabolic mirror is generally more advantageous than a lens-type system, but unfortunately at large aperture ratios the usable field is severely restricted by coma. With an aperture ratio of *f*/3, and a field diameter of only 1°, the spread caused by coma amounts to 37 arc-seconds, and there is an additional spread of 5 arc-seconds produced by astigmatism. Coma increases in direct proportion to the diameter of the field of view, and astigmatism as the square. This means that, close to the axis, astigmatism is insignificant and pure coma predominates, but that away from the axis this is modified by astigmatism.

Nevertheless, a parabolic mirror of an aperture ratio of *f*/8 to *f*/10 is superior to ordinary doublets as regards image sharpness, because mirrors completely lack chromatic aberration. A disadvantage is the fact that the distribution of light in the image discs is asymmetrical, which can produce systematic radial shifts when the discs are being measured.

It should, however, be pointed out that a fully spherical mirror with an aperture ratio of *f*/8 to *f*/10 can be used very satisfactorily. If the aperture stop is placed immediately in front of the mirror there is no advantage over a parabolic mirror, because the spherical mirror has the same faults; indeed spherical aberration is added, which increases the image sizes over the whole of the field. But if the aperture stop is placed at the centre of curvature, the spherical mirror shows no image spread, apart from longitudinal aberration: coma and astigmatism are zero. The image surface then lies on a sphere, which has the same radius as the focal length, and is concentric to the mirror's centre of curvature. The convex side of the image surface is towards the mirror.

The image-diameter at the paraxial focus of a spherical mirror of *f*/8 or *f*/10 amounts to 12.5 or 6.4 arc-seconds, respectively, and the minimum image-diameter is only a quarter of these values, i.e. 3.1 or 1.6 arc-seconds. In practice, therefore, one can obtain sharper images by remaining between those two positions. Under normal circumstances such image-spreads are smaller than the scatter caused within the photographic emulsion itself. In addition, when flat plates are used, the image quality at the edges of the field are better than with a parabolic mirror with a similar focal ratio: all the star images are round, with a symmetrical distribution of light.

But if, using a ring, a circular disc of flat film is pressed into contact with a backing plate that has a curvature corresponding to the spherical image surface, the diffraction discs will be the same size over the whole of the field. The same effect can be achieved by using a steep, plano-convex collector lens placed in front of the photographic plate (plane side towards the plate).

If, however, we want to increase the aperture ratio considerably, the spherical aberration becomes very large, because it varies as the third power of the aperture ratio. At *f*/3 or even *f*/2, the spread at the paraxial focus is 240 or 800 arc-seconds, respectively. The minimum image-diameters are 60 or 200 arc-seconds. With a focal length of 1 m, the paraxial discs would be 1.2 or 4 mm in diameter, and the minimum sizes 0.3 or 1 mm. In such a case, a spherical mirror would also be unusable.

I now wish to show how perfectly sharp images can still be obtained using a spherical mirror with a large aperture ratio.

In order to turn a spherical mirror into a parabolic mirror, the edge has to be flattened, thus giving it a longer radius of curvature. We could also lay a concentric, curved glass shell, with equal thickness throughout, on top of the spherical mirror, and deform one of its surfaces. Now the curvatures have to be reversed, the edge being more steeply curved than the centre. In addition the amount of deformation must be about double that previously required. This is because the deviation is now produced by refraction. In order to obtain the same deviation with refraction as with reflection, the curvature must be about four times as great, but as the light passes through the glass shell twice, only twice as much deformation is required.

But this shell may now be 'bent' optically, so that one face becomes plane, when the other side will have just the deformation curve. In other words, one can equally start by deforming a plane-parallel plate. Optically, when this corrector plate is used, essentially the same results will be obtained as with a parabolic mirror.

Such an appropriately shaped cover plate for a spherical mirror would also have the advantage of providing good protection for the layer of silver. One disadvantage is that about 20% of the light would be lost because of the double passage through the glass.

It is, however, possible to place the corrector plate at another position in the light-path. If it is in front of the image surface, the light passes through the plate only once. The plate must, of course, give twice as much deviation as before. The light-loss is then only 10%.

If we now place the corrector plate at the centre of curvature of the mirror, the conditions are the same as those we had

*Central-Zeitung für Optik und Mechanik, 52. Jahrgang Heft 2.

previously with a spherical mirror and an aperture stop at the centre of curvature, with the exception that now spherical aberration is also removed and, moreover, over the whole of the field. It is therefore possible to use aperture ratios of $f/3$ to $f/2$, and obtain complete freedom from coma, astigmatism, and spherical aberration.

If the angle of incidence of the incoming rays is very large, the projection of the corrector plate is an ellipse, and the deviation is not projected onto the proper areas of the mirror. The correction therefore alters, and indeed becomes an over-correction in the radial direction.

However, large inclinations do not really come into the question, because otherwise the photographic plate would be larger than the free aperture. In practice, one would hardly use photographic plates larger than 1/4 to 1/3 of the aperture, when errors caused by the angle of incidence are still insignificant.

The situation is somewhat different as regards chromatic aberration in the corrector plate. In order to keep this to a minimum, the corrector plate needs to be fashioned so that the centre acts as a weak positive lens, and the edge as a negative one. If the neutral zone lies at 0.866 of the diameter, then chromatism is at a minimum. If the point at which the curvature reverses is at 0.707 of the diameter, the plate's edge thickness is the same as its central thickness. Furthermore, the difference in thickness between the thickest and thinnest parts of the plate are so small – only a few hundredths of a millimetre – that there are no detrimental colour effects. In any case, the effects are far smaller than the secondary spectrum in a corresponding objective.

This chromatism is identical with the so-called 'chromatic difference of spherical aberration'.

If the mirror has the same diameter as the corrector plate, the incident cylinder of rays forming an image at the edge of the field falls eccentrically on the mirror, and a lune-shaped section of the cylinder is lost. The edge of the plate therefore receives somewhat less light. If we want to avoid this, then the mirror should be larger in diameter than the free aperture. It must in fact be larger by twice the diameter of the plate. For a mirror of 50-cm free aperture (diameter of the corrector plate), and 1-m focal length, for a field of 6° the diameter of the photographic plate has to have a diameter of 10.5 cm. The mirror therefore needs to be 71 cm in diameter.

As the foregoing discussion shows, the fast, coma-free, reflecting system described here offers major advantages with respect to speed and error-free imaging. It does, however, require technically perfect manufacture of the corrector plate.

Bergedorf, 1931 January

Bernhard Schmidt

ADDENDUM Mr *B. Schmidt* has manufactured a corrector plate for a spherical mirror available at our observatory, which has a diameter of 44 cm, and a focal length of 62.5 cm. Using this ($f/1.75$) system, Mr Schmidt has taken a series of test plates, with a horizontal arrangement of the makeshift camera tube and using the 60-cm coelostat mirror. Two of the exposures made on the curved film are shown in Plates I and II. It will be seen from these that, despite its great aperture ratio, the reflecting system gives a completely coma-free field of 15° diameter, which would greatly advance many astronomical investigations. It is intended to manufacture a particularly stable camera tube for this reflecting system, and to mount this equatorially, in order to carry out further test exposures.

Bergedorf, 1932 January

R. Schorr

Captions:

Plates obtained with the new Schmidt reflecting system

PLATE I:

Star cluster in Hercules (M13). Exposure taken on the optical axis, duration 40 minutes, enlargement 10 times.

A field in Cygnus. Ten-times enlargement of the corner indicated on Plate II. Distance from the optical axis = 8°. [Plate II is not reproduced here.]

Glossary

Aperture stop The circular boundary of the pencil of light entering a telescope. In a Schmidt telescope, the edge of the corrector plate acts as an aperture stop.

Aspheric surface A surface that deviates from a spherical form. Such surfaces are important in calculating and producing astronomical optics.

Association A loose grouping of stars that formed at the same time (OB-Association, T-Association, etc). Associations are normally found in conjunction with related interstellar matter.

Astronomical Unit (AU) A rounded value for the average distance of the Earth from the Sun (1 AU = 149.6×10^6 km).

Balmer lines The characteristic spectral lines of hydrogen, named after J. Balmer (1825–1898). The Balmer lines arise from transitions occurring from the first energy level above the ground state in the electron shell. In order of decreasing wavelength, they are Hα, Hβ, Hγ, etc.

Bit Abbreviation of 'binary digit'. In information theory, the smallest unit of information, represented by an 'on' or 'off' state, an appropriate pattern of which may be used to encode any information.

Bremsstrahlung Radiation that is produced by the acceleration of electrons in the field of ions or atomic nuclei.

CCD detector Abbreviation for 'Charge-Coupled Device'. A sensitive electronic radiation detector, which is so constructed that it allows a two-dimensional image to be examined.

Chromatic aberration The dependance of the refractive index of a lens on the wavelength of the incident light, which, if uncorrected, gives coloured fringes around the image.

Collapse phase The evolutionary state in which a cloud of diffuse material, or a star, contracts rapidly under the influence of its own gravitational force.

Coma In cometary astronomy, the shell of dust and gas surrounding the nucleus of a comet. In optics, a specific form of aberration, affecting the shape of the image of an object.

Compact galaxy A system of stars with very sharp boundaries and a very small angular diameter. Compact galaxies have a high surface brightness and are surrounded by weak haloes at the most.

Concave Curved inwards. The primary mirror in a reflecting telescope always has a concave surface.

Convex Curved outwards. The outer surface of a sphere is convex.

Cosmogony The theory of the origin and evolution of celestial objects.

Cosmology The science concerned with the study of the state and evolution of the universe as a whole.

Dark cloud A concentration of dust-like material in space between the stars. The light from stars lying behind such clouds – relative to the observer – is weakened by absorption, giving the impression of an area with few or no stars.

Declination An astronomical coordinate, giving the angular distance of an astronomical object from the celestial equator.

Delta-Cephei stars (δ-Cephei stars) Variable stars, named after the prototype object in Cepheus. These stars show periodic variation in magnitude, with a steep rise to maximum and a slower decline. Their periods lie between 1 and 50 days, most being between 3 and 6 days. The period of their variation is closely related to their absolute magnitude.

Dew cap An extension in front of the objective lens of a refractor or the corrector plate of a Schmidt telescope, designed to prevent condensation of moisture in the air occurring on the optical element. In large telescopes there is often an electrical heating system inside the dew cap to promote the effect.

Dielectric layers A dielectric material does not conduct an electrical current, and its special optical characteristics affect its response to incident light. Layers of such a material may be used to achieve a particularly high reflectivity over a specific range of wavelengths.

Diffraction disc The image of a point source that is produced by an optical system at the detector. It is caused by diffraction at the telescope aperture.

Dispersion In optics, the relationship between the refractive index and the wavelength of light. It causes the formation of a spectrum when white light passes through a prism.

Doppler effect Variation in the wavelength of radiation caused by the relative motion of the source of radiation and the receiver. If the distance is decreasing, the wavelength decreases (giving a blue-shift); if it increases, the wavelength increases (producing a red-shift). Named after the physicist Christian Doppler (1803–1853).

Emission-line star Stars with emission lines in their spectra. The lines are formed in extended atmospheres of hot gas around the stars.

Epicycle theory The ancient, geocentric theory of the universe used to explain the occasional apparent retrograde motion of the planets. In this theory, the planets moved in a circle (the epicycle), whose centre moved around the Earth on another circle (the deferent). The epicycle theory was introduced by Apollonius in 200 BC.

Equinox The point and time at which the Sun crosses the celestial equator during its annual apparent path (the ecliptic) around the sky. The vernal equinox defines the position of 0^h Right Ascension, one of the coordinates used in astronomy. The points at which the ecliptic and celestial equator intersect are slowly moving because of precession of the Earth's axis, so the appropriate date (known as the Epoch) must be given when specifying celestial coordinates.

Excitation The transition of an electron in an atom or molecule into a higher-energy state caused by the absorption of a quantum of radiation or by collision with another particle.

Expansion coefficient The relative change in the dimensions of an object as a result of a change in temperature.

Extinction The weakening of radiation through the effect of absorption and scattering.

Extragalactic space The space outside our Galaxy, the Milky Way system.

Focal length In its simplest form, the distance between the centre of an optical surface and the point of intersection of the rays, entering the system as parallel light, and deviated by the surface.

Focal surface The surface at which the rays of light in any optical system are brought to a focus.

Foucault's knife-edge test A method of testing telescope mirrors, developed by the French physicist, Léon Foucault (1819–1868).

Fragmentation The splitting of a large system into smaller subsystems. The concept plays an important part in astronomy in accounting for the formation of galaxies, stars and planets.

Fusion In nuclear physics, the formation of atoms with higher atomic numbers from those with lower ones. Within stars, the fusion of hydrogen to helium nuclei represents an important source of energy.

Galactic equator The line of symmetry of the Milky Way, which defines the reference plane for the galactic coordinate system.

Galaxy The Milky Way system. (Without a capital letter, any independent stellar system.)

Galaxy type The division of galaxies into different morphological types, or classes, according to their physical characteristics: e.g. spiral systems.

Globule A very small interstellar dark cloud, usually with a regular spherical shape.

H I region/ H II region A region of space in which interstellar hydrogen is predominantly neutral (H I region) or ionized (H II region).

Heavy elements Chemical elements whose atomic nuclei generally consist of numerous protons and neutrons. In astronomy, all elements with atomic masses larger than that of helium (2 protons and 2 neutrons) are regarded as being heavy – and are also sometimes known as 'metals'.

Hertzsprung–Russell diagram A diagram representing the relationship between the spectral class of stars and their absolute magnitude, named after Ejnar Hertzsprung (1873–1967) and Henry Norris Russell (1877–1957). The vast majority of stars show a definite correlation between the two values and lie on a narrow band across the diagram, the Main Sequence.

IC *Index Catalogue*, a supplement to the catalogue of clusters and nebulae (the NGC), published at the turn of the century by J. L. Dreyer (1852–1926). Objects are identified by 'IC' followed by a serial number.

Infrared Term given to the range of wavelengths between visible light and sub-millimetre and radio wavelengths.

Interference The interaction of two beams of waves, giving high and low intensities, depending on phase difference. In astronomy, it forms the basis of important methods of improving the resolving power of optical and, particularly, of radio telescopes.

Ion An atom or molecule that has become charged by the removal (or more rarely, by the addition) of one or more electrons.

Ionization The process in which charged particles are produced by the removal of electrons from the shell of an atom or molecule, caused by the injection of energy either from radiation or by collision.

Ionized hydrogen A gas consisting of positive hydrogen ions, i.e. protons.

Jet A narrow, collimated ejection of matter from a system. Jets may be observed in active galaxies, and in stars in the process of formation.

Kelvin (K) The basic unit of temperature. The Kelvin scale has its zero point at absolute zero: 0 K = −273.15°C.

Light-year A popular unit for describing the distance of celestial objects. The distance travelled by light in one year at a speed of approximately 300,000 km/s. One light-year = 9.46×10^{12} km.

Limiting magnitude The faintest magnitude that can be detected with a specific telescope and auxiliary equipment.

Luminosity function The number of stars or stellar systems per unit volume of space and within a unit interval of absolute magnitude.

Magnitude A measure of the brightness of stars. Because of the physiology of the eye, there is a logarithmic relationship with the physical luminous flux. The smaller the value, the brighter the object. A star of magnitude 2 appears fainter than one of magnitude 1. Stars of magnitude 6 are the faintest that can normally be seen with the naked eye. This division of stars into different magnitudes was introduced by the Greek astronomer, Hipparchos (190–125 BC).

Main-Sequence star A star that, when classified by spectral class and absolute brightness, falls within a narrow band running across the Hertzsprung–Russell diagram. This stage of stellar evolution is very stable and long-lasting.

Meridian An imaginary great circle on the celestial sphere, running through the South point on the horizon, the zenith, and the North (and South) Celestial Pole.

Messier Catalogue The catalogue of approximately 100, non-stellar, bright, diffuse or extended objects compiled by Charles Messier in 1776. Each object is designated by the letter 'M' followed by a catalogue number.

Micrometer An auxiliary instrument used on a telescope to measure the precise angular distance between objects on the celestial sphere. In one form, the position of an adjustable thread relative to a fixed one may be read from a suitably calibrated scale.

Mounting The general term for the mechanical arrangement for supporting and moving telescopes. There are various forms, such as the fork mounting, the altazimuth mounting, etc.

Multi-mirror telescope A special form of telescope, with several mirrors on one mounting. The light from the various mirrors is brought to a common focus.

Neutron star Stars, the interior of which consists solely of neutrons, which have been created by the exceptionally high densities encountered. Neutron stars are the end-product of the evolution of certain massive stars. Their average density is approximately 10^{14} g/cm^3. Neutron stars may have a mass several times that of the Sun, but their diameter is only about 20 km.

NGC *New General Catalogue of Clusters and Nebulae*: a catalogue of clusters, nebulae and galaxies, published by J. L. Dreyer (1852–1926) in 1880. Objects are designated by the letters 'NGC' followed by a serial number.

Objective prism A prism for the production of stellar spectra, which is mounted in front of the objective glass of a refractor, or in front of the corrector plate of a Schmidt system.

Optical axis An imaginary line connecting the point of focus on the object side of a lens to that on the image side, or between the vertex and the focus of a concave mirror.

Parallax The angle subtended by a baseline when viewed at a distant point. Using the average diameter of the Earth's orbit as a basis, parallax measurements play a fundamental role in the determination of the distance of celestial objects.

Parsec (pc) The abbreviation for *par*allax *sec*ond, a unit of distance. From a star situated at a distance of 1 pc, the average radius of the Earth's orbit would subtend an angle of 1 second of arc. One pc = 3.09×10^{13} km.

Perihelion The point on the orbit of a planet or comet at which it is nearest to the Sun.

Photoelectric photometer An instrument for measuring the brightness of stars, based on a detector which converts the flux of the incident photons into electrical signals.

Photon The smallest unit of electromagnetic radiation (e.g. of light). Radiation may be absorbed or emitted only in multiples of this basic unit. The energy of a photon is proportional to the frequency of the radiation.

Planetary nebula A bright, circumstellar, gaseous nebula with a regular, frequently ring-like or circular, form. It is created during a late stage of the evolution of certain stars. The name originates from the planet-like appearance when viewed in small telescopes.

Planetology The science of the physical state, formation, and evolution of planets.

Polarization All electromagnetic waves oscillate perpendicular to the direction of propagation. Normally, all possible planes of vibration are represented. If one plane is predominant, the light is said to be polarized. The process that causes this property is known as polarization.

Population A group of objects sharing common characteristics. In astronomy, the term is particularly applied to stars of the same age, the same chemical composition, and with similar other criteria, these groups forming subclasses within a stellar system.

Prime focus The focal point produced by the main (primary) mirror in a telescope system. Frequently, the inclusion of other optical elements gives rise to other, useful focal points, such as the Cassegrain focus or the Coudé focus.

Proper motion The component of a star's motion in space that is at right angles to the line of sight, expressed as an angular distance per unit time. The proper motion of most stars lies below 0.1 arcseconds per year.

Proton An elementary particle with positive charge, and a component of all atomic nuclei. Its mass amounts to 1.67×10^{24} g.

Protostar The product of the fragmentation of an interstellar cloud, and the final stage before the formation of a star. The energy radiated by the protostar was released by the collapse of the cloud, and does not yet involve the fusion of atomic nuclei. The shell of the star is optically thick, and allows only long-wave radiation to escape.

Pulsar A source of radiation at radio wavelengths, the emission of which consists of regular pulses. The periods discovered to date range between 0.001 s and 4.5 s. A few objects are known that are also sources of X-ray emission, and which emit pulses in the optical region. Pulsars are stellar remnants: rapidly rotating neutron stars.

Quantum In its most restricted sense, the smallest unit of radiation. A photon is a quantum of light.

Quantum efficiency The ratio of the quanta registered by a radiation detector to those actually striking it. In photographic emulsions, the quantum efficiency is only about 1%, i.e. only one quantum in one hundred leads to a noticeable photo-chemical reaction in the emulsion.

Quasar An extragalactic object that appears star-like. Quasars have prominent spectra with high red-shifts, but they primarily radiate intense radio emission. Nowadays, it is thought that quasars are the bright, active cores of very distant galaxies.

Radial velocity The component of the motion of a celestial body that lies along the observer's line of sight. It can be measured spectroscopically by means of the Doppler effect.

Radiation pressure Pressure that is exerted by electromagnetic radiation on particles, and the direction of which coincides with the direction of propagation of the light.

Recombination The process in which a free electron is captured by an atomic or molecular ion.

Red giant A star at an advanced stage of evolution, which has a low surface temperature, but which, because of its enormous size, has a high luminosity.

Reflection nebula A dense accumulation of dust particles in interstellar space, which reflect the light of a nearby star and thereby become visible as a diffuse, bright object.

Refraction The change in the direction of a ray of light as it passes from one optical medium to another. Atmospheric refraction is particularly significant and causes a systematic alteration in the measured position of a celestial body as a function of its zenith distance.

Refractive index An optical constant that describes the change in direction of a ray of light as it moves from one medium into another.

Resolving power The smallest angle between two point sources that produces distinct images in a specific telescope.

Right Ascension (RA) An astronomical coordinate, which is measured along the celestial equator. The zero point of Right Ascension is the First Point of Aries – the vernal equinox – the point at which the apparent path of the Sun crosses the celestial equator from South to North.

RR-Lyrae stars A class of variable stars with regular variations of light. The periods of RR-Lyrae stars are short (around 1 day) and their amplitude is about one magnitude.

Seeing A term used to describe the degree of turbulence present in the air, which causes stars to scintillate or 'twinkle'. When large telescopes are being used, poor seeing leads to increased image sizes. Bad seeing also impairs the resolving power of telescopes and their limiting magnitudes.

Solar mass 1.99×10^{30} kg

Solar radius 6.96×10^{5} km

Solar wind A stream of charged particles (mainly protons and electrons) emitted by the Sun.

Spectral class A system of stellar classification on the basis of the spectra. The most common spectral types, in order of decreasing temperature of the stellar surface, are O, B, A, F, G, K, M. A finer sub-division is provided by the addition of a number from 0–9.

Spectral line A discrete, narrow feature in a spectrum, which may be an absorption line (where the intensity is reduced) or an emission line (where the intensity is increased).

Spherical aberration A fault of optical systems, where the focal length is dependent on the distance of the individual rays from the optical axis of the system.

Square degree The unit of surface area on the celestial sphere, expressed in angular dimension. The entire sky consists of 41,250 square degrees.

Stellar wind A continuous stream of charged particles that is emitted by stars, particularly high-luminosity stars.

Strömgren sphere A spherical volume around a star, in which the hydrogen gas is ionized by the stellar radiation. It is named after the astrophysicist, Bengt Strömgren (1908–1987). The size of a Strömgren sphere depends on the density of the gas and the number of ultraviolet quanta emitted by the star per unit time. Only hot stars are therefore able to form significant Strömgren spheres.

Supernova The explosive, final stage in the evolution of massive stars. At maximum, the luminosity reaches extremely high values. At least two different types (arising in different forms of star) may be distinguished from the light-curve and spectrum.

Synchrotron radiation Radiation emitted by electrons that are moving at high speeds in spiral paths around magnetic-field lines.

Transmission The percentage of light transmitted through an optical element. In photographic photometry, the fraction of the incident light that is transmitted by the emulsion is known as the transmission.

Uranostat An optical instrument in which a single plane mirror directs the light from a celestial body into a certain direction, e.g. horizontally. In order to allow for the diurnal motion of the object, the plane mirror has to move at an irregular rate. The device is similar to a coelostat, but the latter uses two plane mirrors.

Vignetting Partial obstruction of the pencil of light entering a telescope by the mechanical components. The degree of vignetting varies according to the position in the field of view, and produces errors in photometric work.

Wavefront An imaginary surface (perpendicular to the direction of propagation) on which waves of light or other radiation have the same phase.

Wolf–Rayet stars Stars with a very high effective temperature (50,000 K to 100,000 K) and luminosity. Wolf–Rayet stars are surrounded by expanding shells of gas, which can be recognized by their characteristic spectral emission lines.

Zonal defect An optical defect that occurs in certain zones of the overall system.

Index

Page references to illustrations are in italic.